# Recurrent Event Modeling Based on the Yule Process

**Mathematical Models and Methods in Reliability Set**

coordinated by

Nikolaos Limnios and Bo Henry Lindqvist

Volume 2

# Recurrent Event Modeling Based on the Yule Process

*Application to Water Network Asset Management*

Yves Le Gat

iSTE

WILEY

First published 2016 in Great Britain and the United States by ISTE Ltd and John Wiley & Sons, Inc.



ISTE Ltd
27-37 St George's Road
London SW19 4EU
UK

www.iste.co.uk

John Wiley & Sons, Inc.
111 River Street
Hoboken, NJ 07030
USA

www.wiley.com



Library of Congress Control Number: 2015955809

British Library Cataloguing-in-Publication Data
A CIP record for this book is available from the British Library
ISBN 978-1-84821-891-8

# Contents

# Preface

The research work presented in this book arises from the involvement of the author in engineering studies of the reliability of drinking water pipes. This type of infrastructure is organized as a network of pipelines, and failures, namely leakage or breakage, tend to occur in an aggregative manner on the same network segments. Building relevant strategies of infrastructure asset management requires, therefore, accurate modeling tools of the repeated failures that can affect some pipes, due to the heavy socioeconomic and environmental consequences of leakage and breakage.

Yves LE GAT
October 2015

# 1

# Introduction

Examples of recurrent failures abound in the literature devoted to the reliability of technical objects, and in many cases, the occurrence rates tend to increase not only with the ageing of the object, but also with the number of past failures. The effect of ageing can be relevantly modeled using the now classical non-homogeneous Poisson process (NHPP), a comprehensive presentation of which can be found in [LAW 87], and a good example of application to drinking water pipe failures in [RØS 00]. In this same context of pipe failures, the PhD work of [EIS 94] emphasizes the critical importance of past failures. The consideration of the dependency of the failure process on its past is not a trivial question, and motivates a theoretical effort which the present book attempts to contribute to.

The basic concept of a *stochastic process* underlies all developments of the present work. A stochastic process must be understood as a function $X()$ of time $t$, each $X(t)$ being considered as a random variable (r.v.).

The stochastic process theory is the *natural* mathematical framework for studying the repetition of random events of the same kind. As presented by [COO 02], this question can be addressed from two alternative perspectives, which are equivalent and respectively consist of modeling:

– either the distribution of successive inter-arrival times;

– or the distribution of the number of events that occur in a given time interval.

The method chosen by [EIS 94] arises from the first approach. The "classical" presentation of [ROS 83] arises from the second approach. The linear extension of the Yule process (called LEYP throughout the rest of the book) aims at building a failure occurrence model that cumulates the advantages of both NHPP and [EIS 94]'s approaches. This involves a theoretical setup, focused on the *counting process* concept, which is to be developed throughout the next two chapters.

A counting process is a particular stochastic process, simply designed to count repeated events, as presented in section 1.2.1.

As this presentation is to have a general scope, the entity subjected to repeated failures will be called a *technical object* or more simply an *object*; this term will be replaced by "water main" or "water pipe" when the context refers more specifically to failures that affect a water network.

## 1.1. Notation

The following mathematical notations will be used throughout this book:

– $\mathbb{N}$ and $\mathbb{N}^*$ respectively denote the sets of natural integers $\{0, 1, 2, \ldots, \infty\}$ and the set of strictly positive natural integers $\{1, 2, \ldots, \infty\}$;

– $\mathbb{R}$, $\mathbb{R}_+$ and $\mathbb{R}_+^*$ are the real sets $]-\infty, +\infty[$, $[0, +\infty[$ and $]0, +\infty[$ ;

– $\mathrm{P}(A)$ and $\mathrm{P}(A \mid B)$ respectively denote the probability of the event $A$, and the conditional probability of $A$ given that the other event $B$ occurs;

– $\mathrm{P}(A \cap B)$ and $\mathrm{P}(A, B)$ equivalently denote the joint probability of events $A$ and $B$ ; $\mathrm{P}\left(\bigcap_j A_j\right)$ more generally stands for the joint probability of events $A_j$;

– $t \in \mathbb{R}_+$ is a positive time variable that stands for the age of a technical object;

– $N(t) \in \mathbb{N}$ is an integer-valued step function that counts the failures;

– $\mathrm{d}N(t) = N(t+\mathrm{d}t) - N(t)$ is the differential of $N(t)$, i.e. $\mathrm{d}N(t) = 1$ whenever a failure occurs within $[t, t+\mathrm{d}t[$, $\mathrm{d}N(t) = 0$ otherwise;

– $\Delta N(t) = N(t) - N(t-)$ stands for the increment of $N(t)$ at $t$;

– $\mathscr{N}_{[a,t[}$ stands for the auto-exciting $\sigma$-algebra generated by the process $N(t)$ within $[a, t[$ ;

– $\mathscr{N}_{t-}$ stands for the auto-exciting $\sigma$-algebra $\mathscr{N}_{[0,t[}$;

– $\mathbf{Z}$ is a vector of failure factor values specific to a given technical object, also called "covariates";

– $\mathscr{F}_{[a,t[} = \mathscr{N}_{[a,t[} \vee \sigma(\mathbf{Z})$ denotes the information on the process $\mathscr{N}_{[a,t[}$ increased by the knowledge of the covariates $\mathbf{Z}$, or more technically the smallest $\sigma$-algebra that contains all events composed with events of $\sigma$-algebras $\mathscr{N}_{[a,t[}$ and $\sigma(\mathbf{Z})$;

– $\lambda(t)$ is a real positive function bounded on any compact interval, and its integral is $\Lambda(t) = \int_0^t \lambda(u)\mathrm{d}u$;

– $\mathrm{E}X$ and $\mathrm{E}(X \mid A)$ respectively denotes the expectation of the random variable (r. v.) $X$ and its conditional expectation given $A$;

– $\mathrm{Var}(X)$ denotes the variance of the r. v. $X$;

– $\mathcal{U}_E$ stands for the uniform distribution on the set $E$;

– $\mathcal{U}_{[0,1]}$ denotes in particular the uniform distribution on interval $[0, 1]$ ;

– $N(\mu, \sigma^2)$ stands for the Gaussian distribution with expectation $\mu$ and variance $\sigma^2$;

– $\mathcal{P}o(\mu)$ is the Poisson distribution with expectation $\mu \in \mathbb{R}_+$;

– $\mathcal{NB}(\theta, p)$ is the negative binomial distribution with two parameters $\theta \in \mathbb{R}_+^*$ and $p \in [0, 1]$;

– $\mathcal{NM}(\theta, (p_j)_{j=1,\dots,n})$ is the negative multinomial distribution with $n + 1$ parameters $\theta \in \mathbb{R}_+^*$ and $p_j \in [0, 1]$;

– $\mathcal{M}(k, (p_j)_{j=1,\dots,n})$ is the multinomial distribution with $n + 1$ parameters $k \in \mathbb{N}^*$ and $p_j \in [0, 1]$, where $\sum_{j=1}^n p_j = 1$;

– $\chi^2(k)$ is the Chi-squared distribution with $k \in \mathbb{N}^*$ degrees of freedom;

– $L(\boldsymbol{\theta})$ stands for the likelihood of a theoretical process with parameter $\boldsymbol{\theta}$ given a sequence of observed events;

– $\prod$ stands for the product integral operator, which plays the same role for products as the integral operator $\int$ plays for sums;

– the indicator function $\mathrm{I}(p)$ of proposition $p$ takes value 1 if $p$ is true, 0 otherwise;

– $s \wedge t$ gives the minimum of scalars $s$ and $t$;

– the operator min() gives the minimum of a collection of values;

– the operator max() gives the maximum of a collection of values.

The calculation lines that build up the proof of a proposition will be closed by a right-justified □ symbol. The text lines that express a remark will be typed in italic and closed by a right-justified △ symbol.

## 1.2. General theoretical framework

The theoretical approach adopted throughout this book builds on two essential reference textbooks. The pioneering *Statistical Models Based on Counting Processes* [AND 93], by P.K. Andersen, Ø. Borgan, R.D. Gill and N. Keiding, emphasizes the power of the concepts of the counting process and intensity function to rigorously process survival data. More recently, *Survival and Event History Analysis* [AAL 08], by O.O. Aalen, Ø. Borgan and H.K. Gjessing, explicitly extends the theoretical framework to properly handle recurrent event data.

### 1.2.1. *The concept of a counting process*

We consider a technical object which is observable in continuous time and is likely to undergo events of interest, also called failures, at random times $T_j$, with $j \in \mathbb{N}$ denoting the rank of the failure. The time variable $t$ is measured since the object considered was put into service, i.e. at $t = 0$, and we will often use the terms "time" and "age" indifferently. By convention, the failure time $T_0$ is not random and fixed at the time the object began to be observed, at age 0 or later. The random variable $T_j$ might then be either the age at the first failure, or at the first observed failure. The time interval within which the object is observed will be denoted by $[a, b]$, with $a \in \mathbb{R}_+, b \in \mathbb{R}_+^*$.

As illustrated in Figure 1.1, the counting process $N(t)$ is a right continuous and left-limited integer-valued function that starts at $N(0) = 0$ and increases by one unit at each $T_j$:

$$\forall t \in \{T_j : j = 1, \ldots, \infty\}, \quad dN(t) = 1$$

$$\forall t \in ]T_j, T_{j+1}[: j = 0, \ldots, \infty, \quad dN(t) = 0$$

It is moreover assumed that at most one failure can occur at a given time, and that the process cannot "explode", i.e. the counting function keeps a finite value at any finite time:

$$\forall t \in \mathbb{R}_+, \begin{cases} P(dN(t) > 1) = 0 \\ P(N(t) < \infty) = 1 \end{cases}$$

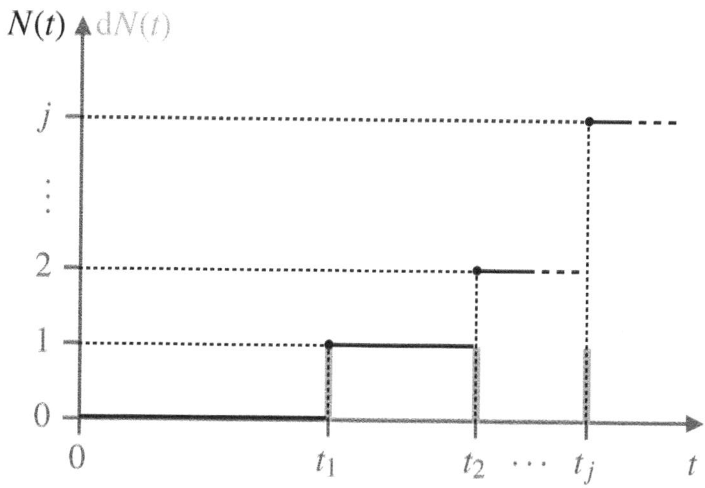


**Figure 1.1.** *Counting process* $N(t)$ *and differential* $dN(t)$

### 1.2.2. *The intensity function of a counting process*

Let $\mathcal{N}_{[a,t[}$ denote the $\sigma$-algebra $\sigma(N(s) - N(a))_{s \in [a,t[}$. Informally called the *past* in [AAL 08], $\mathcal{N}_{[a,t[}$ can be seen as the knowledge available about the process since the beginning of its observation until just before $t$. This information is qualified as *left-truncated* if the failure process is not observed since the object was put into service ($a > 0$), so nothing is known about the process within $[0, a[$.

The intensity function of $N(t)$, which we will denote by $\eta(t)$, can be heuristically defined as the probability density of a one unit jump at $t$, conditional on the past:

$$\mathrm{P}\left(\mathrm{d}N(t) = 1 \mid \mathcal{N}_{[a,t[}\right) = \mathrm{E}\left(\mathrm{d}N(t) \mid \mathcal{N}_{[a,t[}\right)$$

REMARK 1.1.– It is here to be stressed that the main modeling effort presented in this book has consisted of searching for a parametric form as suitable as possible for $\mathrm{E}\left(\mathrm{d}N(t) \mid \mathcal{N}_{[a,t[}\right)$. This conditional expectation assumes an underlying probability distribution for the r.v. $N(t) - N(a) \mid \mathcal{N}_{[a,t[}$, which will generally depend on the parameter denoted by $\boldsymbol{\theta}$; to emphasize the role of $\boldsymbol{\theta}$, the intensity will sometimes be written as $\mathrm{E}_{\boldsymbol{\theta}}\left(\mathrm{d}N(t) \mid \mathcal{N}_{[a,t[}\right)$. △

## 1.3. The non-homogeneous Poisson process

The NHPP model, as presented by [ROS 83], can be defined as:

DEFINITION 1.1.– *The NHPP is defined by the system of equations:*

$$\forall t \in \mathbb{R}_+,\quad \begin{cases} N(0) = 0 \\ \mathrm{E}\left(\mathrm{d}N(t) \mid \mathcal{N}_{t-}\right) = \mathrm{E}\left(\mathrm{d}N(t)\right) = \lambda(t)\mathrm{d}t \end{cases}$$

Pivotal properties of NHPP are:

– the intensity depends on age $t$, hence the term *non-homogeneous*;

– $N(t)$ is Poisson distributed with parameter $\Lambda(t) = \int_0^t \lambda(u)\mathrm{d}u$;

– $N(t)$ is *Markovian*, i.e. its distribution does not depend on the trajectory it took between 0 and $t-$.

The particular intensity function $\lambda(t) = \delta t^{\delta-1} e^{\mathbf{Z}^{\mathrm{T}}\boldsymbol{\beta}}$ is presented by [LAW 87] as tractable for practical use. It is the product of two factors:

– an ageing factor $\delta t^{\delta-1}$, sometimes called *Weibull factor* (see [AAL 08]),

– a scale factor $e^{\mathbf{Z}^{\mathrm{T}}\boldsymbol{\beta}}$, often called *Cox factor*, for it has initially been proposed by [COX 72].

$\mathbf{Z}$ is a vector of explanatory variable values, or *covariates*, which can be either categorical or quantitative, and characterize the technical object or its environment. $\boldsymbol{\beta}$ is a vector of regression coefficients that account for the effects of the covariates on the process intensity. The first components of $\mathbf{Z}$ and $\boldsymbol{\beta}$ are respectively 1 and $\beta_0$, and define the *baseline* intensity, when all other covariate values are 0. The exponential form in the Cox factor make covariates act multiplicatively on the intensity, which makes us qualify this form of NHPP as *proportional hazard model* (PHM), sometimes also called *Cox model*.

## 1.4. The Eisenbeis model

In the model of [EIS 94], which from now will be refered to as the *Eisenbeis model*, the successive inter-event times are random variables $X_j = T_j - T_{j-1}$ defined by $\mathbb{R}_+$, which are indexed by the event occurrence rank $j \in \mathbb{N}$, and follow Weibull distributions with parameters $\mu_j$ and $\delta_j$ that depend on $j$. The cumulative distribution function (CDF) of $X_j$ is written as:

$$\forall x \in \mathbb{R}_+, \forall j \in \mathbb{N}, \mathrm{P}\left(X_j \leq x \mid \mu_j, \delta_j\right) = 1 - \exp\left(-x^{\delta_j} e^{\mu_j}\right)$$

The parameter $\mu_j$ is moreover defined as a linear combination $\mathbf{Z}^{\mathrm{T}}\boldsymbol{\beta}_j$ of explanatory variables (covariates), which can be either categorical or quantitative, and characterize the technical object or its environment. In the technical context of the Eisenbeis model, water mains are characterized by their diameter, length, location under roadway or sidewalk, type of embedding soil, etc. $\boldsymbol{\beta}_j$ is a parameter vector, specific to event rank $j$. As NHPP, this model is thus also a PHM. The components of vectors $\mathbf{Z}$ and $\boldsymbol{\beta}_j$ are indexed by convention from 0 to $q$, where $q$ is the actual number of covariates; a numerical covariate counts indeed for one, whereas a categorical covariate with $m$ possible values counts for $m - 1$ actual covariates (i.e. $m - 1$ indicator variables).

The Eisenbeis model can also be reformulated as the counting process $N(t)$ of the number of events undergone by the object within interval $[0, t]$:

DEFINITION 1.2.– *The Eisenbeis model is defined by the system of equations:*

$$\forall t \in ]T_{j-1}, T_j], \forall j \in \mathbb{N}^*,$$
$$\begin{cases} N(0) = 0 \\ \mathrm{E}\left(\mathrm{d}N(t) \mid N(t-) = j-1\right) = \delta_j (t - T_{N(t-)})^{\delta_j - 1} e^{Z^{\mathrm{T}} \beta_j} \mathrm{d}t \end{cases}$$

*where by convention $T_0 = 0$ at installation of the water main.*

To not have to estimate too many parameters, [EIS 94] proposes to simplify the dependency of $\delta_j$ and $\boldsymbol{\beta}_j$ on $j$ by grouping the values of $j$ into three strata:

– Stratum I for $j \in \{1\}$,

– Stratum II for $j \in \{2, 3, 4\}$

– and Stratum III for $j \in \{5, 6, \ldots\}$,

and by fixing also $\delta_{III} = 1$ in the third stratum.

The respective definitions 1.2 and 1.1 of Eisenbeis and NHPP models highlight an essential difference: the intensity of Eisenbeis model strongly depends on the failure rank, whereas the NHPP is mainly driven by the process age. The counting process based on the Eisenbeis model is additionally *not Markovian*, as its distribution depends on the ages at the previous failures.

## 1.5. Other approaches for water pipe failure modeling

There is an extensive amount of international literature devoted to the modeling of repeated water pipe failures. A relevant overview covering publications since 1979 is given by [KLE 01], more recently completed by [BER 08, BUR 10] and [STC 12]. It is to be noticed that, except for the works focused on inter-failure times, the theoretical framework of stochastic processes is never mentioned. This tendency seems to want to last, since most recent publications, such as [DEB 10] and [YAM 09], promote generalized linear models; [DEB 10] considers the occurrence of at least one failure within time intervals of some years as Poisson distributed, whereas [YAM 09] considers shorter time intervals of some months and the binomial distribution.

## 1.6. Why mobilize the Yule process?

Definition 1.2 of the Eisenbeis model involves an important limitation: estimating parameters by means of observed data is only possible provided that the technical objects are observed since their installation; if observation is oppositely restricted to an age interval $[a, b]$ where $a > 0$, event ranks are unknown and the model cannot therefore be applied.

Practical applications, reported by [LEG 00], have however been carried out to get around the left-truncation issue:

– by consenting to consider that $t = 0$ at the beginning of the observation window;

– and by introducing, in log transforms, the age at the previous failure as well as observed failure ranks as covariates.

Results are interesting on the whole, and show an advantage over NHPP in detecting the water pipes that are the most likely to fail. The Eisenbeis model turns out to be an interesting tool for prioritizing water main renovations or replacements. Predictions of future failure numbers have however always included an embarrassing overestimation tendency. The NHPP, on the other hand, poorly detects water pipes likely to fail, but provides unbiased average predictions. Implementing the Eisenbeis model requires moreover time consuming Monte Carlo computations to get around the impossibility of literally calculating the convolution of Weibull distributions. By contrast, NHPP allows very simple and quick prediction computations.

Investigating the use of the Yule process is then fully justified by the search of a model that would combine the advantages of both Eisenbeis and NHPP models, namely a good ability to detect the objects most likely to undergo future failures, and to provide unbiased and easy to compute predictions. The intensity of the searched process should increase both with age and past failures. The idea to exponentially combine distributed inter-arrival times, the parameter of which depends on the event rank, is mentioned by [LEG 01], who refers to Furry distribution (sometimes also known as the Yule–Furry distribution). The work of [PEL 99] is also to be mentioned, which presents a rigorous solution to handle the Eisenbeis model with observations restricted to age intervals $[a, b]$ that do not start at $a = 0$, by explicitly calculating probabilities $\mathrm{P}(N(b) - N(a) = m \mid N(a) = j)$, and then their expectation over $j$. The idea to mobilize the Yule process is thus greatly indebted to [LEG 01] and [PEL 99]. The theoretical basis of the Yule process is moreover well presented by [ROS 83].

## 1.7. Structure of the book

After this introductory chapter, Chapter 2 will be devoted to preliminary concepts and tools of probability theory; these preliminaries will particularly concern the binomial and multinomial distributions, the negative binomial and multinomial distributions and power series, and their link with the Yule process. Chapter 3 presents the most general form of non-homogeneous birth process (NHBP), insisting particularly on a general formula for the conditional probability of the number of events within a given time interval $[a, b]$ given the number of events that occurred within interval $[0, a[$. This result is then applied in Chapter 4 to the case where the intensity of NHBP linearly depends on the number of past events, which defined the so-called *linear extension of the Yule process* (LEYP); analytical formulas of the negative binomial probability of the number of events within a time interval given increasingly general past observation interval configurations will be established, as well as a negative multinomial generalization for the joint probability of several time intervals (adjacent and non-overlapping). Chapter 5 will establish the likelihood function of a LEYP process given randomly observed sequences of failures, undergone by technical objects characterized by known covariate values; this result is essential to implement an estimation procedure of the LEYP parameters, for which task the interest of the box-constrained Nelder–Mead optimization algorithm will be emphasized. An important extension of LEYP model will then be presented in Chapter 6, aiming at accounting for the selective survival phenomenon; this arises when the LEYP process is not observed since the object installation, and the objects which can be observed are likely to be the most robust among their cohort. This setup involves considering technical objects with a limited service life, and a decommissioning process that depends on the failure process, and at the same time is susceptible to truncate and censor it. This development gives rise to the so-called *LEYP2s*, the likelihood of which is then studied in Chapter 7; this chapter also presents a numerical validation of LEYP2s model parameter estimation procedure. A case study LEYP2s model application, that uses water network data kindly provided by Lausanne (CH) water utility, is presented in Chapter 8, and allows us to check the practical interest of such a statistical tool. Chapter 9 concludes the book.

Chapters 7 and 8 involve some computations, either based on random or actual data, which were all carried out by the author for specific illustration purpose of the book. This whole computational work has been implemented in R scripts [RD 11].

# 2

# Preliminaries

This chapter gathers miscellaneous basic concepts and tools in probability calculus that are useful for further analyzing constructs of the non-homogeneous birth process (NHBP), and as a particular case, the linearly extended Yule process (LEYP). The main covered topics here are the Yule process, the negative binomial distribution and its multinomial generalization, the gamma function and gamma distribution, and the negative binomial and multinomial power series.

Although these fundamental topics are somewhat classical in probability theory, we have considered it useful to present them in detail. Our goal is to familiarize the reader with some analytical handlings indispensable to understand the next chapters, which are not explicitly covered in classical textbooks.

## 2.1. The Yule process and the negative binomial distribution

The so-called *Yule process* was initially presented by G. Udny Yule in his seminal paper [YUL 24], which attempted to formalize the probabilistic bases of evolution theory, and more particularly the temporal increase in the number of taxonomic species in a given genus. The Yule process, also frequently called *pure birth process*, has since become a classical model in stochastic process theory. It was presented by [ROS 83] as a simple model of the temporal growth of a population, initially composed of a single *founder*, and within which individuals reproduce at the constant rate $\lambda > 0$ and have a null mortality rate. The Yule process is then considered by [ROS 83] as the counting process $N(t)$ that represents the population size at time $t$.

DEFINITION 2.1.– *The Yule process with intensity $\lambda \in \mathbb{R}_+$ is defined by the system of equations:*

$$\forall t \in \mathbb{R}_+, \forall j \in \mathbb{N}^*,$$
$$\begin{cases} N(0) = 1 \\ \mathrm{E}\left(\mathrm{d}N(t) \mid \mathcal{N}_{t-}\right) = \mathrm{E}\left(\mathrm{d}N(t) \mid N(t-) = j\right) = j\lambda \mathrm{d}t \end{cases}$$

The process $N(t)$ has two important properties:

– it is a Markovian process, as the probability of a birth between $t$ and $t + \mathrm{d}t$ only depends on the population size *just before t*, denoted by $N(t-)$;

– at most, one birth can occur within an infinitesimal time interval.

As a remarkable consequence, the distribution of $N(t)$ over $\mathbb{N}^*$ is geometric:

$$\forall t \in \mathbb{R}_+, \forall j \in \mathbb{N} : \mathrm{P}\left(N(t) = j + 1\right) = e^{-\lambda t}(1 - e^{-\lambda t})^j \qquad [2.1]$$

This result generalizes to the case of more than one founder, i.e. an initial population size $k \geqslant 1$. The distribution of $N(t)$ over $\mathbb{N} \cap [k, +\infty[$ is obtained by summing the random descendant numbers of the $k$ founders, i.e. by convolving $k$ times the geometric distribution [2.1]. This convolution leads to a *negative binomial distribution* $\mathcal{NB}(k, e^{-\lambda t})$:

$$\forall t \in \mathbb{R}_+, \forall j \in \mathbb{N} \;:\; \mathrm{P}\left(N(t) = j + k\right) = \binom{j+k-1}{k-1} e^{-k\lambda t}(1 - e^{-\lambda t})^j \qquad [2.2]$$

This way of generating a negative binomial distribution is substantially different from the classical distribution presented by [REN 66]. Let the event $A(j, k)$ be defined by $j + k$ random Bernoulli trials, identical and independent with success probability $p$, and along which $j$ failures and $k$ successes are observed, the last trial being successful. The discrete r.v. $J$ takes the value $j \in \mathbb{N}$ when $A(j, k)$ occurs, and its probability function is then:

$$\mathrm{P}\left(J = j\right) = \binom{j+k-1}{k-1} p^k (1 - p)^j \qquad [2.3]$$

This is simply the joint probability of $k$ successes and $j$ failures multiplied by the number of ways to locate the $k - 1$ first successes among the $j + k - 1$

trials that precede the final $k$th success. The negative binomial distribution of $J$ is then denoted by $J \sim \mathcal{NB}(k,p)$.

The definition of the negative binomial distribution $\mathcal{NB}(k,p)$, where $k$ is an integer, generalizes without difficulty to $\mathcal{NB}(\theta,p)$ where $\theta$ is a positive real number. The probability function of integer-valued r.v. $J \in \mathbb{N} \sim \mathcal{NB}(\theta,p)$ is then:

$$\mathrm{P}(J=j) = \frac{\Gamma(\theta+j)}{\Gamma(\theta)j!}p^{\theta}(1-p)^{j} \qquad [2.4]$$

REMARK 2.1.– The form in [2.4] is consistent with [2.3] on account of the property of the Gamma function $\Gamma(.)$:

$$\forall n \in \mathbb{N} \;:\; \Gamma(n+1) = n! \qquad \triangle$$

It is also worth mentioning the formulas to calculate the expectation and variance of the integer-valued r.v. $J \sim \mathcal{NB}(\theta,p)$. To establish them, the following property of the Gamma function $\Gamma(.)$ is used:

$$\forall z \in \mathbb{R}_{+}^{*},\; \Gamma(z+1) = z\Gamma(z) \qquad [2.5]$$

PROPOSITION 2.1.– The expectation of the r.v. $J \sim \mathcal{NB}(\theta,p)$ is:

$$\mathrm{E}(J) = \frac{\theta(1-p)}{p}$$

PROOF.–

We use [2.5] and the change of variable $l = j-1$:

$$\mathrm{E}(J) = \sum_{j=0}^{\infty} j\mathrm{P}(J=j) = \sum_{j=0}^{\infty} j\frac{\Gamma(\theta+j)}{\Gamma(\theta)j!}p^{\theta}(1-p)^{j}$$

$$= \sum_{j=1}^{\infty}(\theta+j-1)\frac{\Gamma(\theta+j-1)}{\Gamma(\theta)(j-1)!}p^{\theta}(1-p)(1-p)^{j-1}$$

$$= \theta(1-p)\sum_{l=0}^{\infty}\frac{\Gamma(\theta+l)}{\Gamma(\theta)l!}p^{\theta}(1-p)^{l} + (1-p)\sum_{l=0}^{\infty} l\frac{\Gamma(\theta+l)}{\Gamma(\theta)l!}p^{\theta}(1-p)^{l}$$

$$= \theta(1-p) + (1-p)\mathrm{E}(J)$$

Which shows that:

$$\mathrm{E}(J) - (1-p)\mathrm{E}(J) = \theta(1-p)$$

And finally:

$$\mathrm{E}(J) = \frac{\theta(1-p)}{p} \square$$

PROPOSITION 2.2.– The variance of the r.v. $J \sim \mathcal{NB}(\theta, p)$ is:

$$\mathrm{Var}(J) = \frac{\theta(1-p)}{p^2}$$

PROOF.–

$$\mathrm{Var}(J) = \sum_{j=0}^{\infty} j^2 \mathrm{P}(J=j) - \mathrm{E}(J)^2$$

We use [2.5] and the change of variable $l = j - 1$:

$$\mathrm{Var}(J) + \mathrm{E}(J)^2 = \sum_{j=0}^{\infty} j^2 \frac{\Gamma(\theta+j)}{\Gamma(\theta)j!} p^\theta (1-p)^j$$

$$= \sum_{j=1}^{\infty} j \frac{\Gamma(\theta+j)}{\Gamma(\theta)(j-1)!} p^\theta (1-p)^j$$

$$= \sum_{j=1}^{\infty} (j-1) \frac{\Gamma(\theta+j)}{\Gamma(\theta)(j-1)!} p^\theta (1-p)^j$$

$$+ \sum_{j=1}^{\infty} \frac{\Gamma(\theta+j)}{\Gamma(\theta)(j-1)!} p^\theta (1-p)^j$$

$$= \sum_{j=1}^{\infty} (j-1)(\theta+j-1) \frac{\Gamma(\theta+j-1)}{\Gamma(\theta)(j-1)!} p^\theta (1-p)^j$$

$$+ \sum_{j=1}^{\infty} (\theta+j-1) \frac{\Gamma(\theta+j-1)}{\Gamma(\theta)(j-1)!} p^\theta (1-p)^j$$

$$= \sum_{j=1}^{\infty} \theta(j-1) \frac{\Gamma(\theta+j-1)}{\Gamma(\theta)(j-1)!} p^{\theta}(1-p)^j$$

$$+ \sum_{j=1}^{\infty} (j-1)^2 \frac{\Gamma(\theta+j-1)}{\Gamma(\theta)(j-1)!} p^{\theta}(1-p)^j$$

$$+ \sum_{j=1}^{\infty} \theta \frac{\Gamma(\theta+j-1)}{\Gamma(\theta)(j-1)!} p^{\theta}(1-p)^j + \sum_{j=1}^{\infty} (j-1) \frac{\Gamma(\theta+j-1)}{\Gamma(\theta)(j-1)!} p^{\theta}(1-p)^j$$

$$= \sum_{l=0}^{\infty} \theta(1-p) l \frac{\Gamma(\theta+l)}{\Gamma(\theta) l!} p^{\theta}(1-p)^l + \sum_{l=0}^{\infty} (1-p) l^2 \frac{\Gamma(\theta+l)}{\Gamma(\theta) l!} p^{\theta}(1-p)^l$$

$$+ \sum_{l=0}^{\infty} \theta(1-p) \frac{\Gamma(\theta+l)}{\Gamma(\theta) l!} p^{\theta}(1-p)^l + \sum_{l=0}^{\infty} (1-p) l \frac{\Gamma(\theta+l)}{\Gamma(\theta) l!} p^{\theta}(1-p)^l$$

$$= \theta(1-p)\mathrm{E}(J) + (1-p)\left(\mathrm{Var}(J) + \mathrm{E}(J)^2\right)$$

$$+ \theta(1-p)(1-p)\mathrm{E}(J)$$

Which shows that:

$$\mathrm{Var}(J) - (1-p)\mathrm{Var}(J)$$
$$= \theta(1-p)\mathrm{E}(J) + (1-p)\mathrm{E}(J)^2 + \theta(1-p) + (1-p)\mathrm{E}(J) - \mathrm{E}(J)^2$$

And, using proposition 2.1:

$$p\mathrm{Var}(J) = p\mathrm{E}(J)^2 + (1-p)\mathrm{E}(J)^2 + p\mathrm{E}(J) + (1-p)\mathrm{E}(J) - \mathrm{E}(J)^2$$

And, finally:

$$\mathrm{Var}(J) = \mathrm{E}(J)/p$$

□

## 2.2. Gamma-mixture of NHPP

The following result are taken from [GRE 20]: if the conditional distribution of r.v. $X$ is Poisson with expectation $\theta$, and $\theta$ is

Gamma-distributed with parameters $\mu, \sigma \in \mathbb{R}_+^*$, then the marginal distribution of $X$ is negative binomial $\mathcal{NB}(\mu, 1/(\sigma+1))$.

PROOF.–

The conditional probability of $X$ is:

$$\mathrm{P}(X = x \mid \theta) = \frac{\theta^x}{x!}e^{-\theta}, \forall x \in \mathbb{N}$$

and the probability density of $\theta \in \mathbb{R}_+ \sim \mathcal{G}(\mu, \sigma)$:

$$f(\theta) = \frac{\theta^{\mu-1}e^{-\theta/\sigma}}{\sigma^\mu \Gamma(\mu)}$$

And the marginal probability of $X$ is then:

$$\begin{aligned}
\mathrm{P}(X = x) &= \int_0^{+\infty} \frac{\theta^x}{x!}e^{-\theta}\frac{\theta^{\mu-1}e^{-\theta/\sigma}}{\sigma^\mu \Gamma(\mu)}\mathrm{d}\theta \\
&= \frac{1}{\Gamma(\mu)x!\sigma^\mu}\int_0^{+\infty} \theta^{x+\mu-1}e^{-\theta(1+\frac{1}{\sigma})}\mathrm{d}\theta \\
&= \frac{\Gamma(\mu+x)}{\Gamma(\mu)x!\sigma^\mu(1+\frac{1}{\sigma})^{\mu+x}} \\
&= \frac{\Gamma(\mu+x)}{\Gamma(\mu)x!}\left(\frac{1}{\sigma+1}\right)^\mu\left(\frac{\sigma}{\sigma+1}\right)^x
\end{aligned}$$

□

REMARK 2.2.– The penultimate equality results from the definition and the following property of the Gamma function (see [ABR 72]):

$$\begin{aligned}
\forall x \in \mathbb{R}_+^*, \forall y \in \mathbb{R}_+^*, \quad \Gamma(x) &= \int_0^{+\infty} t^{x-1}e^{-t}\mathrm{d}t \\
&= y^x \int_0^{+\infty} t^{x-1}e^{-yt}\mathrm{d}t
\end{aligned}$$

△

As a consequence emphasized by [LAW 87], a counting process with intensity function:

$$\mathrm{E}(\mathrm{d}N(t)) = \nu\lambda(t)\mathrm{d}t$$

where $v$ is a Gamma-distributed random factor with expectation 1 and variance $\alpha$:

$$v \in \mathbb{R}_+ \sim \mathcal{G}(\alpha^{-1}, \alpha)$$

has a negative binomial distribution:

$$N(t) \sim \mathcal{NB}\left(\alpha^{-1}, (\alpha\Lambda(t) + 1)^{-1}\right)$$

where $\Lambda(t) = \int_0^t \lambda(u)\mathrm{d}u$.

## 2.3. The negative binomial power series

The proofs of some results will subsequently use the following negative binomial power series:

$$\forall p \in ]0, 1[, \forall \theta \in \mathbb{R}_+^*, \quad \sum_{j=0}^{\infty} \frac{\Gamma(\theta + j)}{\Gamma(\theta) j!} p^j = (1 - p)^{-\theta} \qquad [2.6]$$

PROOF.–

We consider the negative binomial r.v. $J \sim \mathcal{NB}(\theta, 1 - p)$ and sum up its probability function defined by [2.4]:

$$\sum_{j=0}^{\infty} \frac{\Gamma(\theta + j)}{\Gamma(\theta) j!} (1 - p)^{\theta} p^j = 1$$

And, then:

$$\sum_{j=0}^{\infty} \frac{\Gamma(\theta + j)}{\Gamma(\theta) j!} p^j = (1 - p)^{-\theta}$$

□

## 2.4. The negative multinomial distribution

The negative binomial distribution generalizes two random trials with more than two possible outcomes, to give the so-called *negative multinomial distribution*.

We will have to use the negative multinomial distribution in Chapter 6, to handle the joint distribution of several r.v. $D_j, j \in \{1, \ldots, n\}$, with respective probability parameters $p_j$:

$$\bigcap_{j=1}^{n} D_j \sim \mathcal{NM}(\theta, p_1, \ldots, p_n)$$

with:

$$0 < \sum_{j=1}^{n} p_j < 1$$

The joint probability of $D_j$, $j \in \{1, \ldots, n\}$ is given by:

$$\mathrm{P}\left(\bigcap_{j=1}^{n} D_j = d_j\right) = \frac{\Gamma(\theta + \sum_{j=1}^{n} d_j)}{\Gamma(\theta) \prod_{j=1}^{n} d_j!} \left(1 - \sum_{j=1}^{n} p_j\right)^{\theta} \prod_{j=1}^{n} p_j^{d_j} \qquad [2.7]$$

The $\mathcal{NM}$ distribution has two important properties. The first property relates to the distribution of the sum of r.v. the joint distribution of which is $\mathcal{NM}$, and the second property to their joint conditional distribution given their sum.

PROPOSITION 2.3.– If $D_j$, $j \in \{1, \ldots, n\}$ are r.v. such as:

$$\bigcap_{j=1}^{n} D_j \sim \mathcal{NM}(\theta, p_1, \ldots, p_n),$$

then:

$$\sum_{j=1}^{n} D_j \sim \mathcal{NB}\left(\theta, 1 - \sum_{j=1}^{n} p_j\right).$$

PROOF.–

$$\mathrm{P}\left(\sum_{j=1}^{n} D_j = k\right) = \sum_{\sum_{j=1}^{n} d_j = k} \mathrm{P}\left(\bigcap_{j=1}^{n} D_j = d_j\right)$$

$$= \sum_{\sum_{j=1}^{n} d_j = k} \frac{\Gamma(\theta + \sum_{j=1}^{n} d_j)}{\Gamma(\theta) \prod_{j=1}^{n} d_j!} \left(1 - \sum_{j=1}^{n} p_j\right)^{\theta} \prod_{j=1}^{n} p_j^{d_j}$$

$$= \frac{\Gamma(\theta + k)}{\Gamma(\theta) k!} \left(1 - \sum_{j=1}^{n} p_j\right)^{\theta} \sum_{\sum_{j=1}^{n} d_j = k} \frac{k!}{\prod_{j=1}^{n} d_j!} \prod_{j=1}^{n} p_j^{d_j}$$

And using the multinomial theorem:

$$
= \frac{\Gamma(\theta + k)}{\Gamma(\theta)k!} \left(1 - \sum_{j=1}^{n} p_j\right)^{\theta} \left(\sum_{j=1}^{n} p_j\right)^{k}
$$

□

PROPOSITION 2.4.– If $D_j, j \in \{1, \ldots, n\}$ are such as:

$$
\bigcap_{j=1}^{n} D_j \sim \mathcal{NM}(\theta, p_1, \ldots, p_n),
$$

then for any $k \in \mathbb{N}$:

$$
\bigcap_{j=1}^{n} D_j \mid \sum_{j=1}^{n} D_j = k \sim \mathcal{M}\left(k, \frac{p_1}{\sum_{j=1}^{n} p_j}, \ldots, \frac{p_n}{\sum_{j=1}^{n} p_j}\right).
$$

PROOF.–

For any n-tuple $(d_1, \ldots, d_n) \in \mathbb{N}^n$ such as $\sum_{j=1}^{n} d_j = k$:

$$
\begin{aligned}
\mathrm{P}\left(\bigcap_{j=1}^{n} D_j = d_j \mid \sum_{j=1}^{n} D_j = k\right) &= \frac{\mathrm{P}\left(\bigcap_{j=1}^{n} D_j = d_j, \sum_{j=1}^{n} D_j = k\right)}{\mathrm{P}\left(\sum_{j=1}^{n} D_j = k\right)} \\
&= \frac{\mathrm{P}\left(\bigcap_{j=1}^{n} D_j = d_j\right)}{\mathrm{P}\left(\sum_{j=1}^{n} D_j = k\right)} \\
&= \frac{\Gamma(\theta + k)}{\Gamma(\theta) \prod_{j=1}^{n} d_j!} \left(1 - \sum_{j=1}^{n} p_j\right)^{\theta} \prod_{j=1}^{n} {d_j}^{d_j} \\
&\quad \times \left(\frac{\Gamma(\theta + k)}{\Gamma(\theta)k!} \left(1 - \sum_{j=1}^{n} p_j\right)^{\theta} \left(\sum_{j=1}^{n} p_j\right)^{k}\right)^{-1} \\
&= \frac{k!}{\prod_{j=1}^{n} d_j!} \prod_{j=1}^{n} \left(\frac{d_j}{\sum_{j=1}^{n} p_j}\right)^{d_j}
\end{aligned}
$$

□

## 2.5. The negative multinomial power series

To prove some results in Chapter 6, we will need the following multinomial generalization of [2.6]:

$$\sum_{d_1=0}^{\infty} \cdots \sum_{d_n=0}^{\infty} \frac{\Gamma(\theta + \sum_{j=1}^{n} d_j)}{\Gamma(\theta) \prod_{j=1}^{n} d_j!} \prod_{j=1}^{n} p_j^{d_j} = \left(1 - \sum_{j=1}^{n} x_j\right)^{-\theta} \qquad [2.8]$$

The proof outline is the same as for equation [2.6].

# 3

# Non-homogeneous Birth Process

In order to build an extension of the Yule process suitable to model recurrent events, definition 2.1 has to undergo three modifications:

1) as the event of interest is generally speaking a *failure*, no event occurrence is first to be considered before the commissioning of the technical object at $t = 0$. The counting process starts then necessarily with $N(0) = 0$;

2) to account for the loss of reliability as the object ages, the process intensity has vary with time: $\lambda = \lambda(t)$. Such an extension has already been considered by [CHA 02], under the name of *time-dependent Yule process*;

3) it is additionally relevant to consider a dependency of the process on the event rank $j$ more general than the direct proportionality $1 + j$, and the intensity will be proportional to the strictly positive real-valued quantities $\alpha_j$.

The third consideration defines the process usually called *simple birth process* or *pure birth process* in the literature (see [BHA 97] or [SEN 99] for examples). The process we consider in this chapter will then be called *non-homogeneous birth process*, abbreviated as *NHBP*.

DEFINITION 3.1.– *The NHBP is defined by the system of equations:*

$$
\begin{aligned}
&\forall t \in \mathbb{R}_+, \forall j \in \mathbb{N} \; : \\
&\begin{cases} N(0) = 0 \\ \mathrm{E}\left(\mathrm{d}N(t) \mid \mathcal{N}_{t-}\right) = \mathrm{E}\left(\mathrm{d}N(t) \mid N(t-) = j\right) = \alpha_j \lambda(t)\mathrm{d}t \end{cases} \\
&\textit{with}: \quad \alpha_j \in \mathbb{R}_+^*, \quad \forall j,k \in \mathbb{N}: \quad j \neq k \Rightarrow \alpha_j \neq \alpha_k, \quad \textit{and} \quad \alpha_0 = 1
\end{aligned}
$$

As the intensity in definition 2.1, the intensity in definition 3.1, according to the factor $\alpha_j$, confers the Markovian property on process $N(t)$.

According to the Doob–Meyer decomposition theorem (see [AND 93]), the process $N(t)$ can be written as the sum of a predictable process $A(t)$, which is the *model*, and a mean zero martingale $M(t)$, which is the residual the model cannot account for:

$$N(t) = A(t) + M(t)$$

where:

$$A(t) = \int_0^t \alpha_{N(u-)}\lambda(u)\mathrm{d}u$$

## 3.1. NHBP intensity

The probability density of a one unit jump at $t$ of the counting process depends on the $\sigma$-algebra $\mathcal{N}_{t-}$, which is the set of all possible *trajectories* or *paths* of the counting function $N(t)$ between 0 and just before $t$.

In definition 3.1, the process intensity only depends on the value reached by the counting function at $t-$, and the NHBP is then a Markovian process:

$$\mathrm{E}\left(\mathrm{d}N(t) \mid \mathcal{N}_{t-}\right) = \mathrm{E}\left(\mathrm{d}N(t) \mid N(t-) = j\right) = \alpha_j\lambda(t)\mathrm{d}t$$

where:

$$\forall t \in \mathbb{R}_+, \lambda(t) \in \mathbb{R}_+$$

## 3.2. Conditional distribution of the counting process

This section is devoted to establishing an explicit formula for the conditional distribution of the counting process defined in definition 3.1. To that end, we focus on the conditional probability of the number of events that are likely to occur within time interval $[t, t + s]$, with $t, s \in \mathbb{R}_+$, given the number of events that already occurred within $[0, t[$. We use the method presented by [ROS 83] in the case of the NHPP; we fix $t$ and $j$, and introduce the following notation:

$$Q_m(s) = \mathrm{P}\left(N(t + s) - N(t) = m \mid N(t-) = j\right) \qquad [3.1]$$

We first show that $Q_m(s)$ is solution of a linear ordinary differential equation.

PROPOSITION 3.1.– The conditional probability $Q_m(s)$ is solution of the first order linear ordinary differential equation:

$$\forall m \in \mathbb{N}^*, \ \forall s \in \mathbb{R}_+ :$$

$$\mathrm{d}Q_m(s)/\mathrm{d}s + \alpha_{j+m}\lambda(t+s)Q_m(s) = \alpha_{j+m-1}\lambda(t+s)Q_{m-1}(s) \quad [3.2]$$

with the initial condition:

$$Q_0(s) = \exp\left(-\alpha_j[\Lambda(t+s) - \Lambda(t)]\right) \quad [3.3]$$

where:

$$\Lambda(t) = \int_0^t \lambda(u)\mathrm{d}u \quad [3.4]$$

PROOF.–

We first establish the validity of the initial condition [3.3].

For $m = 0$, with $t$ and $j$ fixed, we can write:

$$\begin{aligned} Q_0(s+\mathrm{d}s) &= \mathrm{P}\left(N(t+s+\mathrm{d}s) - N(t) = 0 \mid N(t) = j\right) \\ &= \mathrm{P}\left(N(t+s+\mathrm{d}s) - N(t+s)\right) \\ &= 0, N(t+s) - N(t) = 0 \mid N(t) = j \end{aligned}$$

As $\mathrm{P}\left(N(t) = j\right) \neq 0$ and $\mathrm{P}\left(N(t+s) - N(t) = 0, N(t) = j\right) \neq 0$, we find:

$$\begin{aligned} Q_0(s+\mathrm{d}s) &= \mathrm{P}\left(N(t+s+\mathrm{d}s) - N(t+s)\right) \\ &= 0 \mid N(t+s) - N(t) = 0, N(t) = j \\ &\quad \times \mathrm{P}\left(N(t+s) - N(t) = 0 \mid N(t) = j\right) \end{aligned}$$

As $N(t)$ is a Markovian process:

$$\begin{aligned}&\mathrm{P}\left(N(t+s+\mathrm{d}s)-N(t+s)=0 \mid N(t+s)-N(t)=0, N(t)=j\right)\\&=\mathrm{P}\left(N(t+s+\mathrm{d}s)-N(t+s)=0 \mid N(t+s)=j\right)\end{aligned}$$

So:

$$Q_0(s+\mathrm{d}s)=\left(1-\alpha_j\lambda(t+s)\mathrm{d}s\right)Q_0(s)$$

And also:

$$\begin{aligned}&Q_0(s+\mathrm{d}s)-Q_0(s)=-\alpha_j\lambda(t+s)Q_0(s)\mathrm{d}s\\&\Longrightarrow \quad \mathrm{d}Q_0(s)/Q_0(s)=-\alpha_j\lambda(t+s)\mathrm{d}s \quad \Longrightarrow \quad \mathrm{d}\ln Q_0(s)=-\alpha_j\lambda(t+s)\mathrm{d}s\end{aligned}$$

Then by integration:

$$\int_0^s \mathrm{d}\ln Q_0(u)=-\alpha_j\int_0^s \lambda(t+u)\mathrm{d}u \qquad \text{with:} \quad Q_0(0)=1$$

We finally obtain:

$$Q_0(s)=\exp\left(-\alpha_j[\Lambda(t+s)-\Lambda(t)]\right)$$

We can now prove the validity of proposition [3.2]. For any $m \geq 1, t$ and $j$ being fixed:

$$\begin{aligned}Q_m(s+\mathrm{d}s)&=\mathrm{P}\left(N(t+s+\mathrm{d}s)-N(t)=m \mid N(t)=j\right)\\&=\mathrm{P}\left(N(t+s+\mathrm{d}s)-N(t+s)=0, N(t+s)-N(t)=m \mid N(t)=j\right)\\&\quad+\mathrm{P}\left(N(t+s+\mathrm{d}s)-N(t+s)=1, N(t+s)-N(t)=m-1 \mid N(t)=j\right)\\&=\mathrm{P}\left(N(t+s+\mathrm{d}s)-N(t+s)=0 \mid N(t+s)-N(t)=m, N(t)=j\right)\\&\quad\times\mathrm{P}\left(N(t+s)-N(t)=m \mid N(t)=j\right)\\&\quad+\mathrm{P}\left(N(t+s+\mathrm{d}s)-N(t+s)=1 \mid N(t+s)-N(t)=m-1, N(t)=j\right)\\&\quad\times\mathrm{P}\left(N(t+s)-N(t)=m-1 \mid N(t)=j\right)\end{aligned}$$

Again, from the Markovian property of

$$N(t): \mathrm{P}(N(t+s+\mathrm{d}s) - N(t+s) = 0 \mid N(t+s) - N(t) = m, N(t) = j)$$
$$= \mathrm{P}(N(t+s+\mathrm{d}s) - N(t+s) = 0 \mid N(t+s) = j+m)$$

and:

$$\mathrm{P}(N(t+s+\mathrm{d}s) - N(t+s) = 1 \mid N(t+s) - N(t) = m-1, N(t) = j)$$
$$= \mathrm{P}(N(t+s+\mathrm{d}s) - N(t+s) = 1 \mid N(t+s) = j+m-1)$$

we obtain:

$$Q_m(s+\mathrm{d}s) = \left(1 - \alpha_{j+m}\lambda(t+s)\mathrm{d}s\right) Q_m(s) + \left(\alpha_{j+m-1}\lambda(t+s)\mathrm{d}s\right) Q_{m-1}(s)$$

and then:

$$\mathrm{d}Q_m(s)/\mathrm{d}s + \alpha_{j+m}\lambda(t+s)Q_m(s) = \alpha_{j+m-1}\lambda(t+s)Q_{m-1}(s) \qquad \square$$

REMARK 3.1.– To establish the validity of [3.3], we have used the following property:

A,B,C being such events so as $\mathrm{P}(C) \neq 0$ and $\mathrm{P}(B \cap C) \neq 0$:

$$\mathrm{P}(A \cap B \mid C) = \mathrm{P}(A \mid B \cap C) \times \mathrm{P}(B \mid C)$$

which is easily obtained by considering:

$$\mathrm{P}(A \cap B \mid C) = \frac{\mathrm{P}(A \cap B \cap C)}{\mathrm{P}(C)}$$

and

$$\mathrm{P}(A \mid B \cap C) \times \mathrm{P}(B \mid C) = \frac{\mathrm{P}(A \cap B \cap C)}{\mathrm{P}(B \cap C)} \times \frac{\mathrm{P}(B \cap C)}{\mathrm{P}(C)} \qquad \triangle$$

Proposition 3.1 is a particular case of Kolmogorov differential equations presented by [BHA 97] for discrete value in continuous time Markovian processes.

The analytical form of the general solution of the linear ordinary differential equation [3.2] is suggested by the convolution of exponential distributions with pairwise different parameters, presented in [COX 62].

This leads to the following proposition.

PROPOSITION 3.2.– The conditional probability that an NHBP with intensity defined by definition 3.1 generates $m$ events within time interval $[t, t+s]$ given by $N(t-) = j$ is:

$$\forall m \in \mathbb{N},\ \forall s \in \mathbb{R}_+ : \quad Q_m(s) = \left(\prod_{k=0}^{m-1} \alpha_{j+k}\right) \sum_{k=0}^{m} \frac{e^{-\alpha_{j+k}[\Lambda(t+s)-\Lambda(t)]}}{\prod_{l=0, l\neq k}^{m} (\alpha_{j+l} - \alpha_{j+k})} \qquad [3.5]$$

where $\alpha_0 = 1$ and $j \neq k \Rightarrow \alpha_j \neq \alpha_k$.

PROOF.–

The general solution of the first order linear ordinary differential equation is obtained as follows:

$$v(s) = e^{-\int_0^s \alpha_{j+m}\lambda(t+u)du} \quad \text{and} \quad (s) = \int_0^s \frac{\alpha_{j+m-1}\lambda(t+u)Q_{m-1}(u)}{v(u)} \mathrm{d}u$$

which leads to the solution:

$$\begin{aligned} Q_m(s) &= v(s)w(s) \\ &= e^{\alpha_{j+m}[\Lambda(t)-\Lambda(t+s)]} \int_0^s \alpha_{j+m-1}\lambda(t+u)Q_{m-1}(u)e^{\alpha_{j+m}[\Lambda(t+u)-\Lambda(t)]}\mathrm{d}u \\ &= e^{-\alpha_{j+m}\Lambda(t+s)} \int_0^s \alpha_{j+m-1}\lambda(t+u)Q_{m-1}(u)e^{\alpha_{j+m}\Lambda(t+u)}\mathrm{d}u \end{aligned}$$

The recursion equation [3.5] holds for $m = 1$:

$$\begin{aligned} Q_1(s) &= e^{-\alpha_{j+1}\Lambda(t+s)} \int_0^s \alpha_j\lambda(t+u)Q_0(u)e^{\alpha_{j+1}\Lambda(t+u)}\mathrm{d}u \\ &= e^{-\alpha_{j+1}\Lambda(t+s)} \int_0^s \alpha_j\lambda(t+u)e^{\alpha_{j+1}\Lambda(t+u)-\alpha_j[\Lambda(t+u)-\Lambda(t)]}\mathrm{d}u \end{aligned}$$

$$= e^{-\alpha_{j+1}\Lambda(t+s)+\alpha_j\Lambda(t)} \frac{\alpha_j}{\alpha_{j+1} - \alpha_j} \int_0^s \mathrm{d}[e^{(\alpha_{j+1}-\alpha_j)\Lambda(t+u)}]$$

$$= \alpha_j \left( \frac{e^{\alpha_j[\Lambda(t)-\Lambda(t+s)]}}{\alpha_{j+1} - \alpha_j} + \frac{e^{\alpha_{j+1}[\Lambda(t)-\Lambda(t+s)]}}{\alpha_j - \alpha_{j+1}} \right)$$

which is [3.5] indeed for $m = 1$.

Assuming that [3.5] holds for $m - 1$, we can write:

$$Q_m(s) = e^{-\alpha_{j+m}\Lambda(t+s)} \int_0^s \alpha_{j+m-1}\lambda(t+u) \left( \prod_{k=0}^{m-2} \alpha_{j+k} \right)$$

$$\times \sum_{k=0}^{m-1} \frac{e^{\alpha_{j+k}[\Lambda(t)-\Lambda(t+u)]}}{\prod_{l=0,l\neq k}^{m-1} (\alpha_{j+l} - \alpha_{j+k})} e^{\alpha_{j+m}\Lambda(t+u)} \mathrm{d}u$$

$$= \left( \prod_{k=0}^{m-1} \alpha_{j+k} \right) e^{-\alpha_{j+m}\Lambda(t+s)}$$

$$\times \sum_{k=0}^{m-1} \frac{e^{\alpha_{j+k}\Lambda(t)}}{\prod_{l=0,l\neq k}^{m-1} (\alpha_{j+l} - \alpha_{j+k})} \int_0^s \lambda(t+u) e^{(\alpha_{j+m}-\alpha_{j+k})\Lambda(t+u)} \mathrm{d}u$$

$$= \left( \prod_{k=0}^{m-1} \alpha_{j+k} \right) e^{-\alpha_{j+m}\Lambda(t+s)}$$

$$\times \sum_{k=0}^{m-1} \frac{e^{\alpha_{j+k}\Lambda(t)}}{\prod_{l=0,l\neq k}^{m-1} (\alpha_{j+l} - \alpha_{j+k})} \int_0^s \frac{\mathrm{d}[e^{(\alpha_{j+m}-\alpha_{j+k})\Lambda(t+u)}]}{\alpha_{j+m} - \alpha_{j+k}}$$

$$= \left( \prod_{k=0}^{m-1} \alpha_{j+k} \right) e^{-\alpha_{j+m}\Lambda(t+s)}$$

$$\times \sum_{k=0}^{m-1} \frac{e^{\alpha_{j+k}\Lambda(t)} \left( e^{(\alpha_{j+m}-\alpha_{j+k})\Lambda(t+s)} - e^{(\alpha_{j+m}-\alpha_{j+k})\Lambda(t)} \right)}{\prod_{l=0, l\neq k}^{m} (\alpha_{j+l} - \alpha_{j+k})}$$

$$= \left( \prod_{k=0}^{m-1} \alpha_{j+k} \right) \sum_{k=0}^{m-1} \frac{e^{\alpha_{j+k}[\Lambda(t)-\Lambda(t+s)]} - e^{\alpha_{j+m}[\Lambda(t)-\Lambda(t+s)]}}{\prod_{l=0, l\neq k}^{m} (\alpha_{j+l} - \alpha_{j+k})}$$

$$= \left( \prod_{k=0}^{m-1} \alpha_{j+k} \right) \times \left( \sum_{k=0}^{m-1} \frac{e^{\alpha_{j+k}[\Lambda(t)-\Lambda(t+s)]}}{\prod_{l=0, l\neq k}^{m} (\alpha_{j+l} - \alpha_{j+k})} - e^{\alpha_{j+m}[\Lambda(t)-\Lambda(t+s)]} \sum_{k=0}^{m-1} \frac{1}{\prod_{l=0, l\neq k}^{m} (\alpha_{j+l} - \alpha_{j+k})} \right)$$

Using identity [B.1], which is proven in Appendix B:

$$\sum_{k=0}^{m} \frac{1}{\prod_{l=0, l\neq k}^{m} (\alpha_l - \alpha_k)} = 0$$

we finally obtain:

$$Q_m(s) = \left( \prod_{k=0}^{m-1} \alpha_{j+k} \right) \left( \sum_{k=0}^{m-1} \frac{e^{\alpha_{j+k}[\Lambda(t)-\Lambda(t+s)]}}{\prod_{l=0, l\neq k}^{m} (\alpha_{j+l} - \alpha_{j+k})} + \frac{e^{\alpha_{j+m}[\Lambda(t)-\Lambda(t+s)]}}{\prod_{l=0, l\neq m}^{m} (\alpha_{j+l} - \alpha_{j+m})} \right)$$

$$= \left( \prod_{k=0}^{m-1} \alpha_{j+k} \right) \sum_{k=0}^{m} \frac{e^{\alpha_{j+k}[\Lambda(t)-\Lambda(t+s)]}}{\prod_{l=0, l\neq k}^{m} (\alpha_{j+l} - \alpha_{j+k})}$$

□

REMARK 3.2.– The analytical form [3.5] is similar to equation [3.9] in [SEN 99], which relates to the convolution of exponential distributions with pairwise different parameters. The authors present this result as an extension of the discrete case of the convolution of geometric distributions. Their proof also uses interpolation by Lagrange polynomials, as for proposition B.1 in Appendix B. $\triangle$

# 4

# Linear Extension of the Yule Process

In the previous chapter, we established the general formula [3.5] for the conditional probability of the NHBP. This result is nevertheless not tractable for practical applications. We can somewhat sacrifice the generality to obtain a more practical result, by defining a Markovian process with intensity at $t$ that linearly depends on the value reached by the counting process at $t-$. Such a linear dependency is shown to generate a counting process with a negative binomial distribution.

## 4.1. LEYP intensity

We call *linear extension of the Yule process* (LEYP) the particular case of NHBP, for which the intensity linearly depends on $N(t-)$.

DEFINITION 4.1.– *A LEYP is defined by the following intensity:*

$$\forall t \in \mathbb{R}_+, \forall j \in \mathbb{N}, \alpha \in \mathbb{R}_+^* \ :$$

$$\mathrm{E}\left(\mathrm{d}N(t) \mid N(t-) = j\right) = (1 + \alpha j)\lambda(t)\mathrm{d}t$$

The class of processes defined by definition 4.1 contains as particular case for the non-homogeneous Poisson process, for which $\alpha = 0$, and the time-dependent Yule process, for which $\alpha = 1$. This second case is trivial, whereas setting $\alpha = 0$ raises a non-trivial problem of passage to the limit with respect to the distribution of the counting process, which will be addressed in section 4.3.

## 4.2. Conditional distribution of the LEYP

We are concerned here by the conditional distribution of the number of events likely to occur in a given time interval, which we call *prediction interval* or *prediction window*, given the number of events that happened in a previous time interval, which we call *observation interval* or *observation window*. We will envisage this conditional distribution in increasingly general configurations; this research is not gratuitous but motivated by the construction of the likelihood function that allows us to estimate model parameters from observations, and by the ability to predict numbers of failures in time intervals not necessarily adjacent to the observation window, and even in the case where observations do not start at the commissioning of the technical objects, or may also only be available on disjoint time intervals (intermittent observation).

### 4.2.1. *Distribution of* $N(b) - N(a) \mid N(a-)$

The simplest configuration is illustrated by Figure 4.1, where the observation window $[0, a]$ starts at the commissioning of the technical object, with an adjacent prediction window.

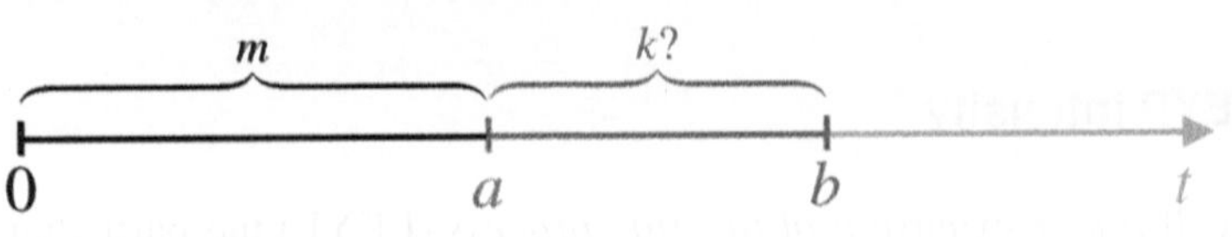


**Figure 4.1.** *Conditional event* $N(b) - N(a)|N(a-)$

In order to simplify the calculations, we introduce the following notation:

$$\mu(t) = e^{\alpha\Lambda(t)} \qquad [4.1]$$

We first prove the following proposition.

PROPOSITION 4.1.– For an LEYP defined by definition 4.1, the number of events likely to occur within observation window $[a, b]$ given by $N(a-) = m$ follows a negative binomial distribution:

$$[N(b) - N(a) \mid N(a-) = m] \sim \mathcal{NB}\left(\alpha^{-1} + m, \frac{\mu(a)}{\mu(b)}\right)$$

PROOF.–

In [3.5], $\alpha_i$ is replaced by $(1 + i\alpha)$:

$$\mathrm{P}(N(b) - N(a) = k \mid N(a-) = m)$$

$$= \left(\prod_{j=0}^{k-1}(1 + (m + j)\alpha)\right)\sum_{j=0}^{k} \frac{\exp\{(1 + (m + j)\alpha)(\Lambda(a) - \Lambda(b))\}}{\prod_{l=0,l\neq j}^{k}\{(1 + (m + l)\alpha) - (1 + (m + j)\alpha)\}}$$

$$= \left(\prod_{j=0}^{k-1}\alpha(\alpha^{-1} + m + j)\right)\sum_{j=0}^{k}\left(\frac{\mu(a)}{\mu(b)}\right)^{\alpha^{-1}+m+j}\left(\prod_{l=0,l\neq j}^{m}\alpha(l - j)\right)^{-1}$$

But:

$$\prod_{j=0}^{k-1}\alpha(\alpha^{-1} + m + j) = \alpha^k\frac{\Gamma(\alpha^{-1} + m + k)}{\Gamma(\alpha^{-1} + m)}$$

and

$$\prod_{l=0,l\neq j}^{k}\alpha(l - j) = \alpha^k(-1)^j j!(k - j)!$$

so:

$$\mathrm{P}(N(b) - N(a) = k \mid N(a-) = m)$$

$$= \frac{\Gamma(\alpha^{-1} + m + k)}{\Gamma(\alpha^{-1} + m)\,k!}\left(\frac{\mu(a)}{\mu(b)}\right)^{\alpha^{-1}+m}\sum_{j=0}^{k}\binom{k}{j}(1)^{k-j}\left(-\frac{\mu(a)}{\mu(b)}\right)^{j}$$

Now, from binomial theorem:

$$\mathrm{P}(N(b) - N(a) = k \mid N(a-) = m)$$

$$= \frac{\Gamma(\alpha^{-1} + m + k)}{\Gamma(\alpha^{-1} + m)\,k!}\left(\frac{\mu(a)}{\mu(b)}\right)^{\alpha^{-1}+m}\left(1 - \frac{\mu(a)}{\mu(b)}\right)^{k}$$

□

### 4.2.2. *Marginal distribution of* $N(t)$

The marginal distribution of $N(t)$ can be directly derived from proposition 4.1 by setting $a = 0, t = b$ and $m = 0$:

PROPOSITION 4.2.– The distribution of LEYP counting process $N(t)$ is negative binomial:

$$\forall t \in \mathbb{R}_+, \quad N(t) \sim \mathcal{NB}(\alpha^{-1}, \mu(t)^{-1})$$

The compensator of LEYP counting process is then:

$$\mathrm{E}N(t) = \frac{\mu(t) - 1}{\alpha}$$

It is interesting to compare this result with the Gamma-mixture of NHPP presented in section 2.2. [ASF 15] compare, both theoretically and practically, the Gamma-mixture of NHPP and the LEYP, considered as NHPP *extensions*, respectively called *heterogeneous* and *dynamic*.

### 4.2.3. *Marginal distribution of* $N(b) - N(a)$

The marginal distribution of $N(b) - N(a)$ can be derived from proposition 4.1:

PROPOSITION 4.3.–

$$\forall a, b \in \mathbb{R}_+, \text{ such as } a < b, \quad N(b) - N(a) \sim \mathcal{NB}\left(\alpha^{-1}, \frac{1}{\mu(b) - \mu(a) + 1}\right)$$

PROOF.–

We can first calculate $\mathrm{P}(N(b) - N(a) = k \mid N(a-) = m)$, and then use [2.6]:

$$\mathrm{P}(N(b) - N(a) = k)$$

$$= \sum_{m=0}^{\infty} \mathrm{P}(N(b) - N(a) = k \mid N(a-) = m)\, \mathrm{P}(N(a-) = m)$$

$$= \sum_{m=0}^{\infty} \frac{\Gamma(\alpha^{-1} + m + k)}{\Gamma(\alpha^{-1} + m)\, k!} \left(\frac{\mu(a)}{\mu(b)}\right)^{\alpha^{-1}+m} \left(\frac{\mu(b) - \mu(a)}{\mu(b)}\right)^{k} \frac{\Gamma(\alpha^{-1} + m)}{\Gamma(\alpha^{-1})\, m!}$$

$$\left(\frac{1}{\mu(a)}\right)^{\alpha^{-1}} \left(\frac{\mu(a)-1}{\mu(a)}\right)^{m}$$

$$= \frac{\Gamma(\alpha^{-1}+k)}{\Gamma(\alpha^{-1})k!} \left(\frac{1}{\mu(b)}\right)^{\alpha^{-1}} \left(\frac{\mu(b)-\mu(a)}{\mu(b)}\right)^{k} \sum_{m=0}^{\infty} \frac{\Gamma(\alpha^{-1}+m+k)}{\Gamma(\alpha^{-1}+k)m!} \left(\frac{\mu(a)-1}{\mu(b)}\right)^{m}$$

$$= \frac{\Gamma(\alpha^{-1}+k)}{\Gamma(\alpha^{-1})k!} \left(\frac{1}{\mu(b)}\right)^{\alpha^{-1}} \left(\frac{\mu(b)-\mu(a)}{\mu(b)}\right)^{k} \left(1-\frac{\mu(a)-1}{\mu(b)}\right)^{-(\alpha^{-1}+k)}$$

$$= \frac{\Gamma(\alpha^{-1}+k)}{\Gamma(\alpha^{-1})k!} \left(\frac{1}{\mu(b)-\mu(a)+1}\right)^{\alpha^{-1}} \left(\frac{\mu(b)-\mu(a)}{\mu(b)-\mu(a)+1}\right)^{k}$$

□

### 4.2.4. *Conditional distribution of* $N(a-)$ *given* $N(b)-N(a)$

The following proposition will be useful later on:

PROPOSITION 4.4.– The conditional probability of $N(a-)$, given as $N(b) - N(a)$ $(0 < a < b)$ is negative binomial:

$$[N(a-) \mid N(b)-N(a)=m] \sim \mathcal{NB}\left(\alpha^{-1}+m, \frac{\mu(b)-\mu(a)+1}{\mu(b)}\right)$$

PROOF.–

$$\mathrm{P}(N(a-)=j \mid N(b)-N(a)=m)$$

$$= \frac{\mathrm{P}(N(a-)=j, N(b)-N(a)=m)}{\mathrm{P}(N(b)-N(a)=m)}$$

$$= \frac{\mathrm{P}(N(b)-N(a)=m \mid N(a-)=j)\,\mathrm{P}(N(a-)=j)}{\mathrm{P}(N(b)-N(a)=m)}$$

Using propositions 4.1, 4.2 and 4.3:

$$= \frac{\Gamma(\alpha^{-1}+j+m)}{\Gamma(\alpha^{-1}+j)m!} \left(\frac{\mu(a)}{\mu(b)}\right)^{\alpha^{-1}+j} \left(\frac{\mu(b)-\mu(a)}{\mu(b)}\right)^{m} \frac{\Gamma(\alpha^{-1}+j)}{\Gamma(\alpha^{-1})\,j!}$$

$$\left(\frac{1}{\mu(a)}\right)^{\alpha^{-1}} \left(\frac{\mu(a)-1}{\mu(a)}\right)^{j}$$

$$\times \left( \frac{\Gamma(\alpha^{-1}+m)}{\Gamma(\alpha^{-1})\, m!} \left( \frac{1}{\mu(b)-\mu(a)+1} \right)^{\alpha^{-1}} \left( \frac{\mu(b)-\mu(a)}{\mu(b)-\mu(a)+1} \right)^{m} \right)^{-1}$$

$$= \frac{\Gamma(\alpha^{-1}+j+m)}{\Gamma(\alpha^{-1}+m)\, j!} \left( \frac{\mu(b)-\mu(a)+1}{\mu(b)} \right)^{\alpha^{-1}+m} \left( \frac{\mu(a)-1}{\mu(b)} \right)^{j} \qquad \square$$

### 4.2.5. *Conditional distribution of* $N(c) - N(b)$ *given* $N(b-) - N(a)$

The first generalization of proposition 4.1 considers an observation interval $[a, b]$ that does not start at the commissioning of the technical object (i.e. $a$ may not be zero), and an adjacent prediction interval $[b, c]$ as illustrated by Figure 4.2.

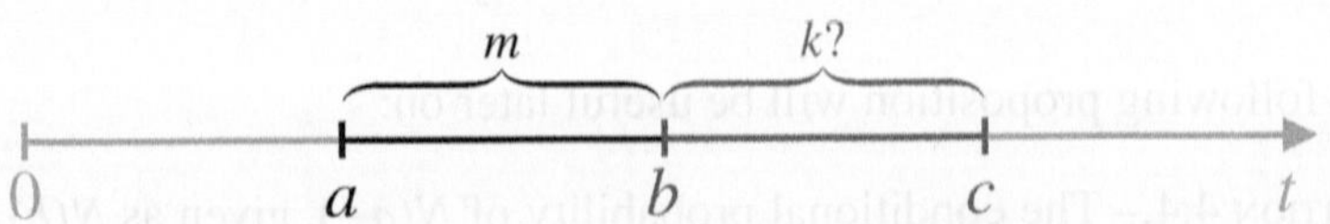


**Figure 4.2.** *Conditional event* $N(c) - N(b)|N(b-) - N(a)$

PROPOSITION 4.5.– The conditional distribution of $N(c)-N(b)$, given as $N(b-)-N(a)$, with $0 \leqslant a < b < c$, is negative binomial:

$$[N(c) - N(b) \mid N(b-) - N(a) = m] \sim \mathcal{NB}\left(\alpha^{-1} + m, \frac{\mu(b)-\mu(a)+1}{\mu(c)-\mu(a)+1}\right)$$

PROOF.–

According to the total probability formula:

$$\mathrm{P}\,(N(c) - N(b) = k \mid N(b-) - N(a) = m)$$

$$= \sum_{j=0}^{\infty} \mathrm{P}\,(N(c) - N(b) = k \mid N(b-) - N(a) = m, N(a-) = j)$$

$$\times \mathrm{P}\,(N(a-) = j \mid N(b-) - N(a) = m)$$

$$= \sum_{j=0}^{\infty} \mathrm{P}\,(N(c) - N(b) = k \mid N(b-) = j + m)\,\mathrm{P}\,(N(a-) = j \mid N(b-) - N(a) = m)$$

Using proposition 4.4:

$$\mathrm{P}\left(N(a-) = j \mid N(b-) - N(a) = m\right)$$

$$= \frac{\Gamma(\alpha^{-1} + j + m)}{\Gamma(\alpha^{-1} + m) j!} \left(\frac{\mu(b) - \mu(a) + 1}{\mu(b)}\right)^{\alpha^{-1}+m} \left(\frac{\mu(a) - 1}{\mu(b)}\right)^{j}$$

Then:

$$\mathrm{P}\left(N(c) - N(b) = k \mid N(b-) - N(a) = m\right)$$

$$= \sum_{j=0}^{\infty} \frac{\Gamma(\alpha^{-1} + j + m + k)}{\Gamma(\alpha^{-1} + j + m) k!} \left(\frac{\mu(b)}{\mu(c)}\right)^{\alpha^{-1}+j+m} \left(\frac{\mu(c) - \mu(b)}{\mu(c)}\right)^{k}$$

$$\times \frac{\Gamma(\alpha^{-1} + j + m)}{\Gamma(\alpha^{-1} + m) j!} \left(\frac{\mu(b) - \mu(a) + 1}{\mu(b)}\right)^{\alpha^{-1}+m} \left(\frac{\mu(a) - 1}{\mu(b)}\right)^{j}$$

$$= \frac{\Gamma(\alpha^{-1} + m + k)}{\Gamma(\alpha^{-1} + m) k!} \left(\frac{\mu(b) - \mu(a) + 1}{\mu(c)}\right)^{\alpha^{-1}+m} \left(\frac{\mu(c) - \mu(b)}{\mu(c)}\right)^{k}$$

$$\times \sum_{j=0}^{\infty} \frac{\Gamma(\alpha^{-1} + m + k + j)}{\Gamma(\alpha^{-1} + m + k) j!} \left(\frac{\mu(a) - 1}{\mu(c)}\right)^{j}$$

Using equation [2.6]:

$$= \frac{\Gamma(\alpha^{-1} + m + k)}{\Gamma(\alpha^{-1} + m) k!} \left(\frac{\mu(b) - \mu(a) + 1}{\mu(c)}\right)^{\alpha^{-1}+m} \left(\frac{\mu(c) - \mu(b)}{\mu(c)}\right)^{k}$$

$$\left(1 - \frac{\mu(a) - 1}{\mu(c)}\right)^{-(\alpha^{-1}+m+k)}$$

$$= \frac{\Gamma(\alpha^{-1} + m + k)}{\Gamma(\alpha^{-1} + m) k!} \left(\frac{\mu(b) - \mu(a) + 1}{\mu(c) - \mu(a) + 1}\right)^{\alpha^{-1}+m} \left(\frac{\mu(c) - \mu(b)}{\mu(c) - \mu(a) + 1}\right)^{k}$$

□

### 4.2.6. *Distribution of* $N(b-) - N(a)$ *given* $N(c) - N(b)$

The following generalization of proposition 4.4 will be useful later on:

PROPOSITION 4.6.– If a LEYP is defined by definition 4.1, the conditional probability of $N(b-) - N(a)$ given as $N(c) - N(b)$, with $0 < a < b < c$, is negative binomial:

$$[N(b-) - N(a) \mid N(c) - N(b) = k] \sim \mathcal{NB}\left(\alpha^{-1} + k, \frac{\mu(c) - \mu(b) + 1}{\mu(c) - \mu(a) + 1}\right)$$

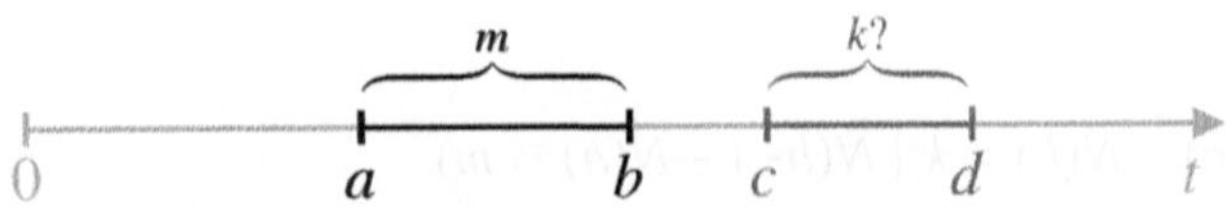


**Figure 4.3.** *Conditional event* $N(d) - N(c)|N(b) - N(a)$

PROOF.–

We proceed as for the proof of proposition 4.4:

$$\begin{aligned}
&\mathrm{P}(N(b-) - N(a) = m \mid N(c) - N(b) = k) \\
&= \frac{\mathrm{P}(N(c) - N(b) = k \mid N(b-) - N(a) = m)\,\mathrm{P}(N(b-) - N(a) = m)}{\mathrm{P}(N(c) - N(b) = k)} \\
&= \frac{\Gamma(\alpha^{-1} + m + k)}{\Gamma(\alpha^{-1} + m)k!}\left(\frac{\mu(b) - \mu(a) + 1}{\mu(c) - \mu(a) + 1}\right)^{\alpha^{-1}+m}\left(\frac{\mu(c) - \mu(b)}{\mu(c) - \mu(a) + 1}\right)^{k} \\
&\times \frac{\Gamma(\alpha^{-1} + m)}{\Gamma(\alpha^{-1})m!}\left(\frac{1}{\mu(b) - \mu(a) + 1}\right)^{\alpha^{-1}}\left(\frac{\mu(b) - \mu(a)}{\mu(b) - \mu(a) + 1}\right)^{m} \\
&\times \left(\frac{\Gamma(\alpha^{-1} + k)}{\Gamma(\alpha^{-1})k!}\left(\frac{1}{\mu(c) - \mu(b) + 1}\right)^{\alpha^{-1}}\left(\frac{\mu(c) - \mu(b)}{\mu(c) - \mu(b) + 1}\right)^{k}\right)^{-1} \\
&= \frac{\Gamma(\alpha^{-1} + m + k)}{\Gamma(\alpha^{-1} + k)m!}\left(\frac{\mu(c) - \mu(b) + 1}{\mu(c) - \mu(a) + 1}\right)^{\alpha^{-1}+k}\left(\frac{\mu(b) - \mu(a)}{\mu(c) - \mu(a) + 1}\right)^{m}
\end{aligned}$$

□

### 4.2.7. *Distribution of* $N(d) - N(c)$ *given* $N(b) - N(a)$

The following proposition provides a further generalization, where the observation and prediction intervals $[a, b]$ and $[c, d]$ are not adjacent, as illustrated by Figure 4.3:

PROPOSITION 4.7.– If a LEYP is defined by [4.1], the conditional distribution of $N(d) - N(c)$ given as $N(b) - N(a) = m$, with $0 < a < b < c < d$, is negative binomial:

$$[N(d) - N(c) \mid N(b) - N(a) = m]$$
$$\sim \mathcal{NB}\left(\alpha^{-1} + m, \frac{\mu(b) - \mu(a) + 1}{\mu(d) - \mu(c) + \mu(b) - \mu(a) + 1}\right)$$

PROOF.–

Using both total probability formula and proposition 4.5

$$\mathrm{P}\,(N(d) - N(c) = k \mid N(b) - N(a) = m)$$
$$= \sum_{j=0}^{\infty} \mathrm{P}\,(N(d) - N(c) = k \mid N(b) - N(a) = m, N(c-) - N(b+) = j)$$
$$\times \mathrm{P}\,(N(c-) - N(b+) = j \mid N(b) - N(a) = m)$$
$$= \sum_{j=0}^{\infty} \mathrm{P}\,(N(d) - N(c) = k \mid N(c-) - N(a) = m + j)$$
$$\times \mathrm{P}\,(N(c-) - N(b+) = j \mid N(b) - N(a) = m)$$
$$= \sum_{j=0}^{\infty} \frac{\Gamma(\alpha^{-1} + m + j + k)}{\Gamma(\alpha^{-1} + m + j)k!} \left(\frac{\mu(c) - \mu(a) + 1}{\mu(d) - \mu(a) + 1}\right)^{\alpha^{-1}+m+j} \left(\frac{\mu(d) - \mu(c)}{\mu(d) - \mu(a) + 1}\right)^{k}$$
$$\times \frac{\Gamma(\alpha^{-1} + m + j)}{\Gamma(\alpha^{-1} + m)j!} \left(\frac{\mu(b) - \mu(a) + 1}{\mu(c) - \mu(a) + 1}\right)^{\alpha^{-1}+m} \left(\frac{\mu(c) - \mu(b)}{\mu(c) - \mu(a) + 1}\right)^{j}$$
$$= \frac{\Gamma(\alpha^{-1} + m + k)}{\Gamma(\alpha^{-1} + m)k!} \left[\left(\frac{\mu(c) - \mu(a) + 1}{\mu(d) - \mu(a) + 1}\right)\left(\frac{\mu(b) - \mu(a) + 1}{\mu(c) - \mu(a) + 1}\right)\right]^{\alpha^{-1}+m} \left(\frac{\mu(d) - \mu(c)}{\mu(d) - \mu(a) + 1}\right)^{k}$$
$$\times \sum_{j=0}^{\infty} \frac{\Gamma(\alpha^{-1} + m + j + k)}{\Gamma(\alpha^{-1} + m + k)j!} \left[\left(\frac{\mu(c) - \mu(a) + 1}{\mu(d) - \mu(a) + 1}\right)\left(\frac{\mu(c) - \mu(b)}{\mu(c) - \mu(a) + 1}\right)\right]^{j}$$

And, using the series [2.6]

$$= \frac{\Gamma(\alpha^{-1} + m + k)}{\Gamma(\alpha^{-1} + m)k!} \left(\frac{\mu(b) - \mu(a) + 1}{\mu(d) - \mu(a) + 1}\right)^{\alpha^{-1}+m} \left(\frac{\mu(d) - \mu(c)}{\mu(d) - \mu(a) + 1}\right)^{k}$$

$$\times \left(1 - \frac{\mu(c) - \mu(b)}{\mu(d) - \mu(a) + 1}\right)^{-(\alpha^{-1}+m+k)}$$

$$= \frac{\Gamma(\alpha^{-1} + m + k)}{\Gamma(\alpha^{-1} + m)k!} \left(\frac{\mu(b) - \mu(a) + 1}{\mu(d) - \mu(c) + \mu(b) - \mu(a) + 1}\right)^{\alpha^{-1}+m}$$

$$\times \left(\frac{\mu(d) - \mu(c)}{\mu(d) - \mu(c) + \mu(b) - \mu(a) + 1}\right)^{k}$$

□

This result is practically important regarding the application of the LEYP model to actual failure data, with the aim of helping decisions in matters of infrastructure renovation.

### 4.3. Limiting distribution when $\alpha$ tends to 0+

As mentioned in section 4.1, setting $\alpha = 0$ in definition 4.1 of the LEYP intensity straightforwardly gives the NHPP intensity (see definition 1.1). However, $\alpha$ cannot be set to 0 in the negative binomial distributions of section 4.2, as their first parameter contains the term $\alpha^{-1}$. In this section, we will attempt to overcome this difficulty by investigating the limiting behavior of the distributions when $\alpha$ tends to 0+.

We can show on this matter that the distribution of the LEYP counting process in proposition 4.2 tends to a Poisson distribution when $\alpha$ tends to 0+.

PROPOSITION 4.8.– The negative binomial distribution $\mathcal{NB}(\alpha^{-1}, e^{-\alpha\Lambda(t)})$ tends to the Poisson distribution $\mathcal{P}o(\Lambda(t))$ when $\alpha$ tends to 0+.

PROOF.–

On the one hand, we first notice that:

$$\alpha^m\Gamma(\alpha^{-1} + m)/\Gamma(\alpha^{-1}) = \alpha^m \frac{1}{\alpha}\left(\frac{1}{\alpha} + 1\right)\left(\frac{1}{\alpha} + 2\right)\cdots\left(\frac{1}{\alpha} + m - 1\right)$$

$$= \frac{\alpha}{\alpha}\left(\frac{\alpha}{\alpha} + \alpha\right)\left(\frac{\alpha}{\alpha} + 2\alpha\right)\cdots\left(\frac{\alpha}{\alpha} + (m-1)\alpha\right)$$

$$= (1 + \alpha)(1 + 2\alpha)\cdots(1 + (m-1)\alpha)$$

which leads to:

$$\lim_{\alpha \to 0+} \alpha^m \Gamma(\alpha^{-1} + m)/\Gamma(\alpha^{-1}) = 1$$

On the other hand, we have by definition:

$$\forall x \in \mathbb{R}, \quad e^x = \sum_{j=0}^{\infty} x^j / j!$$

and it follows that:

$$\begin{aligned} 1 - e^{-\alpha\Lambda(t)} &= 1 - \left(1 + [-\alpha\Lambda(t)] + [-\alpha\Lambda(t)]^2/2! + [-\alpha\Lambda(t)]^3/3! + \cdots\right) \\ &= \alpha\Lambda(t) - [\alpha\Lambda(t)]^2/2! + [\alpha\Lambda(t)]^3/3! - \cdots \end{aligned}$$

then:

$$\frac{1 - e^{-\alpha\Lambda(t)}}{\alpha} = \Lambda(t) - \alpha\Lambda(t)^2/2! + \alpha^2\Lambda(t)^3/3! - \cdots$$

and so:

$$\lim_{\alpha \to 0+} \frac{1 - e^{-\alpha\Lambda(t)}}{\alpha} = \Lambda(t)$$

The above can now be applied to the distribution $\mathcal{NB}(\alpha^{-1}, e^{-\alpha\Lambda(t)})$ :

$$\begin{aligned} \lim_{\alpha \to 0+} \mathrm{P}(N(t) = m) &= \lim_{\alpha \to 0+} \frac{\Gamma(\alpha^{-1} + m)}{\Gamma(\alpha^{-1})m!} \left(e^{-\alpha\Lambda(t)}\right)^{\alpha^{-1}} \left(1 - e^{-\alpha\Lambda(t)}\right)^m \\ &= \lim_{\alpha \to 0+} \frac{\Gamma(\alpha^{-1} + m)}{\Gamma(\alpha^{-1})} \alpha^m \frac{e^{-\Lambda(t)}}{m!} \left(\frac{1 - e^{-\alpha\Lambda(t)}}{\alpha}\right)^m \\ &= \frac{\Lambda(t)^m}{m!} e^{-\Lambda(t)} \end{aligned}$$

which is indeed the probability function of the distribution $\mathcal{Po}(\Lambda(t))$. □

## 4.4. Partition of an interval

We will now address a topic which is not only in itself theoretically interesting, but also practically in view of the study of selective survival, which will be addressed later in Chapter 6.

We consider now an age interval $[a_0, a_n[$, with $0 \leqslant a_0 < a_n$, which is partitioned into $n$ adjacent non-overlapping subintervals $[a_{j-1}, a_j[, j \in \{1, \ldots, n\}$, as illustrated by Figure 4.4. To lighten the notations, the random number of events $N(a_j) - N(a_{j-1})$ likely to occur within intervals $[a_{j-1}, a_j[$ will be denoted by $D_j$, and its possible realizations $d_j$; $\mu(a_j)$ will also be abbreviated as $\mu_j$, and $\mu(a_j) - \mu(a_{j-1})$ as $\Delta\mu_j$.

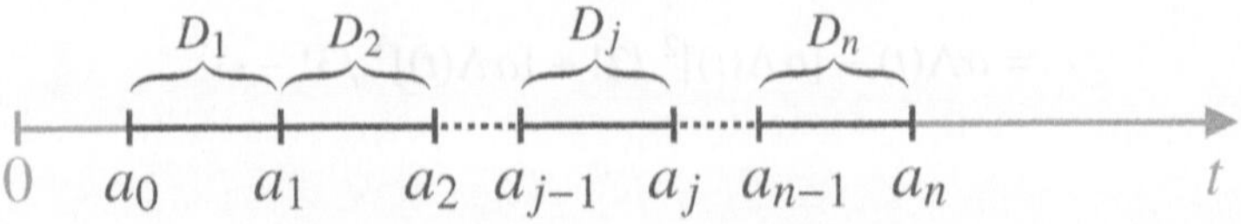


**Figure 4.4.** *Partition of interval* $[a_0, a_n[$

We will now establish the two following results:

– the joint probability of r.v. $\bigcap_{j=1}^{n} D_j$ is *negative multinomial*, which generalizes the negative binomial distribution of any single r.v. $D_j$;

– the conditional joint probability of r.v. $\bigcap_{j=1}^{n} D_j$ given their sum is *multinomial*.

PROPOSITION 4.9.– The joint distribution of r.v. $\bigcap_{j=1}^{n} D_j$ is negative multinomial:

$$\bigcap_{j=1}^{n} D_j \sim NM\left(\alpha^{-1}, \left(\frac{\Delta\mu_j}{1 + \sum_{i=1}^{n} \Delta\mu_i}, j \in \{1, \ldots, n\}\right)\right)$$

PROOF.–

We proceed by induction, and first prove that the proposition holds for $n = 2$:

$$\begin{aligned} &\mathrm{P}(D_1 = d_1, D_2 = d_2) \\ &= \mathrm{P}(D_2 = d_2 \mid D_1 = d_1)\,\mathrm{P}(D_1 = d_1) \end{aligned}$$

$$= \frac{\Gamma(\alpha^{-1} + \sum_{j=1}^{2} d_j)}{\Gamma(\alpha^{-1} + d_1) d_2!} \left( \frac{1 + \Delta\mu_1}{1 + \sum_{j=1}^{2} \Delta\mu_j} \right)^{\alpha^{-1} + d_1} \left( \frac{\Delta\mu_2}{1 + \sum_{j=1}^{2} \Delta\mu_j} \right)^{d_2}$$

$$\times \frac{\Gamma(\alpha^{-1} + d_1)}{\Gamma(\alpha^{-1}) d_1!} \left( \frac{1}{1 + \Delta\mu_1} \right)^{\alpha^{-1}} \left( \frac{\Delta\mu_1}{1 + \Delta\mu_1} \right)^{d_1}$$

$$= \frac{\Gamma(\alpha^{-1} + \sum_{j=1}^{2} d_j)}{\Gamma(\alpha^{-1}) \prod_{j=1}^{2} d_j!} \left( \frac{1}{1 + \sum_{j=1}^{2} \Delta\mu_j} \right)^{\alpha^{-1}} \prod_{j=1}^{2} \left( \frac{\Delta\mu_j}{1 + \sum_{i=1}^{2} \Delta\mu_i} \right)^{d_j}$$

We assume now that the proposition holds for a given $n$:

$$\mathrm{P}\left( \bigcap_{j=1}^{n} D_j = d_j \right) = \frac{\Gamma(\alpha^{-1} + \sum_{j=1}^{n} d_j)}{\Gamma(\alpha^{-1}) \prod_{j=1}^{n} d_j!} \left( \frac{1}{1 + \sum_{j=1}^{n} \Delta\mu_j} \right)^{\alpha^{-1}} \prod_{j=1}^{n} \left( \frac{\Delta\mu_j}{1 + \sum_{i=1}^{n} \Delta\mu_i} \right)^{d_j}$$

Then for $n + 1$:

$$\mathrm{P}\left( \bigcap_{j=1}^{n+1} D_j = d_j \right)$$

$$= \mathrm{P}\left( \bigcap_{j=1}^{n} D_j = d_j, D_{n+1} = d_{n+1} \right)$$

$$= \mathrm{P}\left( D_{n+1} = d_{n+1} \mid \bigcap_{j=1}^{n} D_j = d_j \right) \mathrm{P}\left( \bigcap_{j=1}^{n} D_j = d_j \right)$$

But from the Markovian property:

$$\mathrm{P}\left( D_{n+1} = d_{n+1} \mid \bigcap_{j=1}^{n} D_j = d_j \right)$$

$$= \mathrm{P}\left( D_{n+1} = d_{n+1} \mid \sum_{j=1}^{n} D_j = \sum_{j=1}^{n} d_j \right)$$

And then, using proposition 4.5 and the induction hypothesis:

$$\mathrm{P}\left(\bigcap_{j=1}^{n+1} D_j = d_j\right)$$

$$= \frac{\Gamma(\alpha^{-1} + \sum_{j=1}^{n} d_j)}{\Gamma(\alpha^{-1} + \sum_{j=1}^{n} d_j) d_{n+1}!} \left(\frac{1 + \sum_{j=1}^{n} \Delta\mu_j}{1 + \sum_{j=1}^{n+1} \Delta\mu_j}\right)^{\alpha^{-1} + \sum_{j=1}^{n} d_j} \left(\frac{\Delta\mu_{n+1}}{1 + \sum_{j=1}^{n+1} \Delta\mu_j}\right)^{d_{n+1}}$$

$$\times \frac{\Gamma(\alpha^{-1} + \sum_{j=1}^{n} d_j)}{\Gamma(\alpha^{-1}) \prod_{j=1}^{n} d_j!} \left(\frac{1}{1 + \sum_{j=1}^{n} \Delta\mu_j}\right)^{\alpha^{-1}} \prod_{j=1}^{n} \left(\frac{\Delta\mu_j}{1 + \sum_{j=1}^{n} \Delta\mu_j}\right)^{d_j}$$

$$= \frac{\Gamma(\alpha^{-1} + \sum_{j=1}^{n+1} d_j)}{\Gamma(\alpha^{-1}) \prod_{j=1}^{n+1} d_j!} \left(\frac{1}{1 + \sum_{j=1}^{n+1} \Delta\mu_j}\right)^{\alpha^{-1}} \prod_{j=1}^{n+1} \left(\frac{\Delta\mu_j}{1 + \sum_{j=1}^{n+1} \Delta\mu_j}\right)^{d_j} \quad \square$$

PROPOSITION 4.10.– The conditional joint distribution of r.v. $\bigcap_{j=1}^{n} D_j$ given their sum $\sum_{j=1}^{n} D_j$ is multinomial:

$$\bigcap_{j=1}^{n} D_j \mid \sum_{j=1}^{n} D_j = k \sim \mathcal{M}\left(k, \left(\frac{\Delta\mu_j}{\sum_{i=1}^{n} \Delta\mu_i}, j \in \{1, \ldots, n\}\right)\right)$$

PROOF.–

The proof is a straight forward consequence of proposition 2.4 with:

$$p_j = \frac{\Delta\mu_j}{1 + \sum_{i=1}^{n} \Delta\mu_i}$$

which yields:

$$\frac{p_j}{\sum_{i=1}^{n} p_i} = \frac{\Delta\mu_j}{\sum_{i=1}^{n} \Delta\mu_i} \quad \square$$

## 4.5. Generalization to any subset of a partition

We consider the partition of interval $[a_0, a_n[$ into $n$ subintervals:

$$[a_0, a_n[ = \bigcup_{i=1}^{n} [a_{i-1}, a_i[$$

We name $I$ the complete index set $\{1, \ldots, n\}$, and partition it into two non-empty and non-overlapping index sets $J \subset I$ and $K \subset I$:

$$I = J \cup K, \quad J \neq \emptyset, \quad K \neq \emptyset, \quad J \cap K = \emptyset$$

We will now generalize proposition 4.9 to any subset of the partition of an interval, and keep to this end the same notations as in previous section 4.4. It is important to note that the intervals indexed by $J$ are now not necessarily contiguous.

PROPOSITION 4.11.– The joint distribution of r.v. $\bigcap_{j\in J} D_j$ is negative multinomial:

$$\bigcap_{j\in J} D_j \sim \mathcal{NM}\left(\alpha^{-1}, \left(\frac{\Delta\mu_j}{1 + \sum_{l\in J} \Delta\mu_l}, j \in J\right)\right)$$

PROOF.–

The proof is based on the law of total probability:

$$\begin{aligned}
&\mathrm{P}\left(\bigcap_{j\in J} D_j = d_j\right) \\
&= \sum_{d_k=0,k\in K}^{\infty} \mathrm{P}\left(\bigcap_{j\in J} D_j = d_j \mid \bigcap_{k\in K} D_k = d_k\right) \mathrm{P}\left(\bigcap_{k\in K} D_k = d_k\right) \\
&= \frac{\mathrm{P}\left(\bigcap_{j\in J} D_j = d_j, \bigcap_{k\in K} D_k = d_k\right)}{\mathrm{P}\left(\bigcap_{k\in K} D_k = d_k\right)} \mathrm{P}\left(\bigcap_{k\in K} D_k = d_k\right) \\
&= \sum_{d_k=0,k\in K}^{\infty} \mathrm{P}\left(\bigcap_{i\in J\cup K} D_i = d_i\right) \\
&= \sum_{d_k=0,k\in K}^{\infty} \frac{\Gamma(\alpha^{-1} + \sum_{i\in J\cup K} d_i)}{\Gamma(\alpha^{-1}) \prod_{i\in J\cup K} d_i!} \frac{\prod_{i\in J\cup K} (\Delta\mu_i)^{d_i}}{\left(1 + \sum_{i\in J\cup K} \Delta\mu_j\right)^{\alpha^{-1} + \sum_{i\in J\cup K} d_i}} \\
&= \frac{\Gamma(\alpha^{-1} + \sum_{j\in J} d_j)}{\Gamma(\alpha^{-1}) \prod_{j\in J} d_j!} \frac{\prod_{j\in J} \left(\Delta\mu_j\right)^{d_j}}{(1 + \sum_{i\in J\cup K} \Delta\mu_i)^{\alpha^{-1} + \sum_{j\in J} d_j}}
\end{aligned}$$

$$\times \sum_{d_k=0,k\in K}^{\infty} \frac{\Gamma(\alpha^{-1} + \sum_{i\in J\cup K} d_i)}{\Gamma(\alpha^{-1} + \sum_{j\in J} d_j) \prod_{k\in K} d_k!} \prod_{k\in K} \left(\frac{\Delta\mu_k}{1 + \sum_{i\in J\cup K} \Delta\mu_i}\right)^{d_k}$$

And using multinomial power series [2.8] yields:

$$= \frac{\Gamma(\alpha^{-1} + \sum_{j\in J} d_j)}{\Gamma(\alpha^{-1}) \prod_{j\in J} d_j!} \frac{\prod_{j\in J} \left(\Delta\mu_j\right)^{d_j}}{(1 + \sum_{i\in J\cup K} \Delta\mu_i)^{\alpha^{-1}+\sum_{j\in J} d_j}} \left(1 - \frac{\sum_{j\in J} \Delta\mu_j}{1 + \sum_{i\in J\cup K} \Delta\mu_i}\right)^{-(\alpha^{-1}+\sum_{j\in J} d_j)}$$

$$= \frac{\Gamma(\alpha^{-1} + \sum_{j\in J} d_j)}{\Gamma(\alpha^{-1}) \prod_{j\in J} d_j!} \frac{\prod_{j\in J} \left(\Delta\mu_j\right)^{d_j}}{\left(1 + \sum_{j\in J} \Delta\mu_j\right)^{\alpha^{-1}+\sum_{j\in J} d_j}}$$

□

A further step of generalization can be reached by considering the conditional distribution of joint events $D_j$ indexed on any index set $J$ given as joint events $D_k$ indexed on a subset of the complement of $J$ in $I$.

PROPOSITION 4.12.– The conditional joint distribution of $\bigcap_{j\in J} D_j$ given as $\bigcap_{k\in K} D_k$, with $K \subset I \setminus J$, is negative multinomial:

$$\bigcap_{j\in J} D_j \mid \bigcap_{k\in K} D_k = d_k \sim NM\left(\alpha^{-1} + \sum_{k\in K} d_k, \left(\frac{\Delta\mu_j}{1 + \sum_{i\in J\cup K} \Delta\mu_i}, j \in J\right)\right)$$

PROOF.–

$$P\left(\bigcap_{j\in J} D_j = d_j \mid \bigcap_{k\in K} D_k = d_k\right)$$

$$= \frac{P(\bigcap_{i\in J\cup K} D_i = d_i)}{P(\bigcap_{k\in K} D_k = d_k)}$$

Applying proposition 4.11 both to numerator and denominator yields:

$$= \frac{\Gamma(\alpha^{-1} + \sum_{i\in J\cup K} d_i)}{\Gamma(\alpha^{-1}) \prod_{i\in J\cup K} d_i!} \frac{\prod_{i\in J\cup K} (\Delta\mu_i)^{d_i}}{(1 + \sum_{i\in J\cup K} \Delta\mu_i)^{\alpha^{-1}+\sum_{i\in J\cup K} d_i}}$$

$$\times \left(\frac{\Gamma(\alpha^{-1} + \sum_{k\in K} d_k)}{\Gamma(\alpha^{-1}) \prod_{k\in K} d_k!} \frac{\prod_{k\in K} (\Delta\mu_k)^{d_k}}{(1 + \sum_{k\in K} \Delta\mu_k)^{\alpha^{-1}+\sum_{k\in K} d_k}}\right)^{-1}$$

Rearranging terms and simplifying yields:

$$= \frac{\Gamma(\alpha^{-1} + \sum_{i\in J\cup K} d_i)}{\Gamma(\alpha^{-1} + \sum_{k\in K} d_k) \prod_{j\in J} d_j!} \frac{(1 + \sum_{k\in K} \Delta\mu_k)^{\alpha^{-1}+\sum_{k\in K} d_k}}{(1 + \sum_{i\in J\cup K} \Delta\mu_i)^{\alpha^{-1}+\sum_{i\in J\cup K} d_i}} \prod_{j\in J} \left(\Delta\mu_j\right)^{d_j}$$

□

## 4.6. Discontinuous observation interval

A direct application of proposition 4.12 concerns the case of an observation window composed of non-adjacent intervals, as illustrated by Figure 4.5. The following proposition will be practically useful when, for some reasons either structural or accidental, the observation process is intermittent.

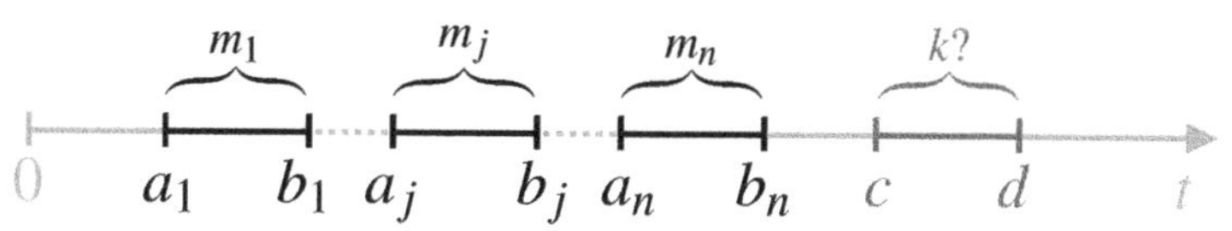


**Figure 4.5.** *Conditional event $N(d) - N(c)$ given as $\bigcap_j N(b_j) - N(a_j)$*

PROPOSITION 4.13.– The conditional distribution of $N(d) - N(c)$, given by $\bigcap_{j=1}^{n} N(b_j - N(a_j) = m_j$, is negative binomial:

$$N(d) - N(c) \mid \bigcap_{j=1}^{n} N(b_j - N(a_j) = m_j$$

$$\sim NM\left(\alpha^{-1} + \sum_{j=1}^{n} m_j, \frac{1 + \sum_{j=1}^{n} \mu(b_j) - \mu(a_j)}{\mu(d) - \mu(c) + 1 + \sum_{j=1}^{n} \mu(b_j) - \mu(a_j)}\right)$$

PROOF.–

The proof follows from a straightforward application of proposition 4.12.

□

# 5

# LEYP Likelihood and Inference

We will now address the estimation of LEYP parameters from observation data related to a sample of technical objects. As already mentioned in section 4.2, the estimation procedure developed in this chapter is required to be usable with failure historics which are not available since the commissioning of the objects. This is particularly the case for water mains, which were extensively installed in most of the European urban areas before 1940, and the maintenance archives of which cover a period posterior to 1985 or often later.

## 5.1. LEYP likelihood

We consider first a single technical object observed within $[a, b]$, where $a \in \mathbb{R}_+$, $b \in \mathbb{R}_+^*$ and $a < b$, and which underwent $m \in \mathbb{N}$ failures at times $t_1 < \ldots < t_j < \ldots < t_m$. The method to infer LEYP parameter estimates from these observations consists of building the likelihood function of the intensity parameters given $t_j$, and searching for the parameter values at which the likelihood reaches its maximum.

An intuitive construction of the likelihood function is presented by [SAM 94] in the NHPP case. It consists of calculating the product of the probabilities that no failure occurs within intervals $]t_j, t_{j+1}[$, and of the limits for $h \to 0+$ of the probabilities that one failure occurs within each $[t_j, t_j + h[$. Applying this to the LEYP framework involves considering all the probabilities as conditional given $N(t_j)$. The developments carried out in Chapter 4 allow us to circumvent the difficulty that arises from the lack of information within $[0, a[$, and then about the value of $N(a)$ and the actual rank of the successive observed failures. The likelihood construct proposed here is well adapted to handle left-truncated failure data.

In order to rigorously construct the likelihood function, we rely upon the general concept of *likelihood in the sense of Jacod*, presented in [AND 93]. The theoretical LEYP likelihood of parameter $\boldsymbol{\theta}$, given a failure sequence, is formally defined as the following product integral:

$$L(\boldsymbol{\theta}) = \prod_{t\in[a,b]} \mathrm{E}\left(\mathrm{d}N(t) \mid \mathcal{N}_{[a,t[}\right)^{\Delta N(t)} \left(1 - \mathrm{E}\left(\mathrm{d}N(t) \mid \mathcal{N}_{[a,t[}\right)\right)^{1-\Delta N(t)} \qquad [5.1]$$

with:

$$\Delta N(t) = N(t) - N(t-)$$

Using the product integral to define the likelihood of a counting process involves considering the observation window $[a, b]$ as an infinite countable sequence of infinitesimal intervals, each of which undergoes a Bernouilli trial with the alternative outcomes:

– either a failure with probability $\mathrm{E}\left(\mathrm{d}N(t) \mid \mathcal{N}_{[a,t[}\right)$;

– or no failure with complementary probability $1 - \mathrm{E}\left(\mathrm{d}N(t) \mid \mathcal{N}_{[a,t[}\right)$.

The basic concept of product integration used in the present context is presented in Appendix A.

In order to provide an explicit analytical form for [5.1], we first have to establish the following proposition.

PROPOSITION 5.1.–

$$\mathrm{E}\left(\mathrm{d}N(t) \mid \mathcal{N}_{[a,t[}\right)$$

$$= \left(\alpha^{-1} + (N(t-) - N(a))\right)\mathrm{d}\ln\left(\mu(t) - \mu(a) + 1\right)$$

$$= \left(1 + (N(t-) - N(a))\alpha\right)\frac{\mu(t)\lambda(t)\mathrm{d}t}{\mu(t) - \mu(a) + 1}$$

PROOF.–

As a result of definition 4.1, and using both proposition 4.5 and the continuity of function $\mu()$:

$$\mathrm{E}\left(\mathrm{d}N(t) \mid \mathcal{N}_{[a,t[}\right) = \mathrm{P}\left(N(t + \mathrm{d}t) - N(t) = 1 \mid N(t-) - N(a)\right)$$

$$= \left(\alpha^{-1} + N(t-) - N(a)\right)\left(\frac{\mu(t) - \mu(a) + 1}{\mu(t + \mathrm{d}t) - \mu(a) + 1}\right)^{\alpha^{-1}+N(t-)-N(a)}$$

$$\left(\frac{\mu(t + \mathrm{d}t) - \mu(t)}{\mu(t + \mathrm{d}t) - \mu(a) + 1}\right)$$

$$= \left(\alpha^{-1} + N(t-) - N(a)\right)\frac{\mathrm{d}\mu(t)}{\mu(t) - \mu(a) + 1}$$

$$= \left(\alpha^{-1} + (N(t-) - N(a))\right)\mathrm{d}\ln\left(\mu(t) - \mu(a) + 1\right)$$

NOTE.– $\mathrm{d}\mu(t)$ can also be written as $\alpha\mu(t)\lambda(t)\mathrm{d}t$. □

We can now establish the following proposition that provides a closed analytical form for the LEYP likelihood.

PROPOSITION 5.2.– The likelihood of the theoretical process LEYP with parameter $\boldsymbol{\theta}$ given a sequence of observed failures is:

$$L(\boldsymbol{\theta}) = \alpha^m \frac{\Gamma(\alpha^{-1} + m)}{\Gamma(\alpha^{-1})} \frac{\prod_{j=1}^m \mu(t_j)\lambda(t_j)}{(\mu(b) - \mu(a) + 1)^{\alpha^{-1}+m}}$$

PROOF.–

As they are a finite number $m$ of jumps within $[a, b]$:

$$\prod_{t\in[a,b]} \mathrm{E}\left(\mathrm{d}N(t) \mid \mathcal{N}_{[a,t[}\right)^{\Delta N(t)} \left(1 - \mathrm{E}\left(\mathrm{d}N(t) \mid \mathcal{N}_{[a,t[}\right)\right)^{1-\Delta N(t)}$$

$$= \prod_{j=1}^{m} \mathrm{E}\left(\mathrm{d}N(t_j) \mid \mathcal{N}_{[a,t_j[}\right) \prod_{j=0}^{m} \prod_{t\in]t_j,t_{j+1}[} \left(1 - \mathrm{E}\left(\mathrm{d}N(t) \mid \mathcal{N}_{[a,t[}\right)\right)$$

Ignoring $(\mathrm{d}t)^m$ (see [AND 93]), the left side can be written:

$$\prod_{j=1}^{m} \mathrm{E}\left(\mathrm{d}N(t_j) \mid \mathcal{N}_{[a,t_j[}\right)$$

$$= \prod_{j=1}^{m} \mathrm{E}\left(\mathrm{d}N(t_j) \mid N(t_j-) - N(a) = j - 1\right)$$

$$= \prod_{j=1}^{m} (1 + (j-1)\alpha) \frac{\mu(t_j)\lambda(t_j)}{\mu(t_j) - \mu(a) + 1}$$

$$= \alpha^m \frac{\Gamma(\alpha^{-1} + m)}{\Gamma(\alpha^{-1})} \prod_{j=1}^{m} \frac{\mu(t_j)\lambda(t_j)}{\mu(t_j) - \mu(a) + 1}$$

Using the product integral property $\prod(1 - \mathrm{d}X) = \exp\left(-\int \mathrm{d}X\right)$ (see Appendix A), the right side can be written:

$$\prod_{j=0}^{m} \prod_{t \in ]t_j, t_{j+1}[} (1 - \mathrm{E}(\mathrm{d}N(t) \mid N(t-) - N(a) = j))$$

$$= \prod_{j=0}^{m} \exp\left( - \int_{t_j}^{t_{j+1}} \left(\alpha^{-1} + j\right) \frac{\mathrm{d}\mu(t)}{\mu(t) - \mu(a) + 1} \right)$$

$$= \prod_{j=0}^{m} \exp\left( - \int_{t_j}^{t_{j+1}} \mathrm{d} \ln \left(\mu(t) - \mu(a) + 1\right)^{\alpha^{-1}+j} \right)$$

$$= \prod_{j=0}^{m} \left( \frac{\mu(t_j) - \mu(a) + 1}{\mu(t_{j+1}) - \mu(a) + 1} \right)^{\alpha^{-1}+j}$$

$$= \frac{\prod_{j=1}^{m} \mu(t_j) - \mu(a) + 1}{(\mu(b) - \mu(a) + 1)^{\alpha^{-1}+m}}$$

Simplifying by $\prod_{j=1}^{m} \left(\mu(t_j) - \mu(a) + 1\right)$ completes the proof. □

For a *random* $n$-sample of technical objects assumed to fail each $m_i$ times within $[a_i, b_i]$, $i \in \{1, \ldots, n\}$, independently of each other, the global likelihood is the product of the individual likelihoods:

$$L(\theta) = \prod_{i=1}^{n} L_i(\theta)$$

In practice, the log-likelihood is preferentially used, as it involves a sum calculation numerically more tractable than a product. The log-likelihood for

$n$ technical objects is written as:

$$\ln L(\boldsymbol{\theta}) = \sum_{i=1}^{n} \Big( m_i \ln \alpha + \ln \Gamma\left(\alpha^{-1} + m_i\right) - \ln \Gamma\left(\alpha^{-1}\right)$$

$$- \left(\alpha^{-1} + m_i\right) \ln \left(\mu(b_i) - \mu(a_i) + 1\right) + \sum_{j=1}^{m_i} \ln \lambda(t_{ij}) + \alpha \Lambda(t_{ij}) \Big) \qquad [5.2]$$

## 5.2. LEYP parameter estimation

### 5.2.1. *Maximum likelihood estimator*

The theoretical outline of maximum likelihood (ML) estimation assumes the observed technical objects as randomly drawn from a theoretical infinite population, the failure process of which follows a LEYP with parameter $\boldsymbol{\vartheta}$. This theoretical parameter is *not random* but unknown. The ML estimator of $\boldsymbol{\vartheta}$ is a random vector:

$$\hat{\boldsymbol{\theta}}_n = \arg_{\boldsymbol{\theta}} \max \ln L(\boldsymbol{\theta})$$

which we conjecture as being asymptotically (1) unbiased, (2) efficient and (3) normally distributed:

$$(1) \quad \lim_{n \to \infty} \mathrm{E}\left(\hat{\boldsymbol{\theta}}_n\right) = \boldsymbol{\vartheta}$$

$$(2) \quad \lim_{n \to \infty} \mathrm{Var}\left(\hat{\boldsymbol{\theta}}_n\right) = \left( - \frac{\partial^2 \ln L(\boldsymbol{\theta})}{\partial \theta^2} \right)_{\boldsymbol{\vartheta}}^{-1}$$

$$(3) \quad \lim_{n \to \infty} \left( - \frac{\partial^2 \ln L(\boldsymbol{\theta})}{\partial \theta^2} \right)_{\boldsymbol{\vartheta}}^{\frac{1}{2}} \left(\hat{\boldsymbol{\theta}}_n - \boldsymbol{\vartheta}\right) \sim \mathcal{N}(\mathbf{0}, \mathbf{1})$$

where $\mathbf{0}$ and $\mathbf{1}$ are respectively the null vector and the identity matrix of same dimension as $\boldsymbol{\theta}$.

We are not able to show that these properties hold in the LEYP case, and comply with the theoretical ML properties. The asymptotic behavior of the ML estimator results from regularity conditions of the likelihood function, presented by [RAO 73] or [COX 74], which seem difficult to assert with the analytical form [5.2]. We graphically check that the shape of the likelihood function allows us to search its maximum, using to that end randomly simulated datasets of several thousands of technical objects, assumed to follow known failure process with realistic failure rates. This verification work will be presented in Chapter 6 in the more general LEYP2s framework.

### 5.2.2. *Null hypothesis of parameter estimates*

Property (2) stated in section 5.2.1 allows us to estimate the covariance matrix of the parameter estimates $\hat{\theta}$:

$$\widehat{\mathrm{Var}}(\hat{\theta}) = \left(-\frac{\partial^2 \ln L(\theta)}{\partial \theta^2}\right)^{-1}_{\hat{\theta}} \qquad [5.3]$$

This makes it possible to carry out a *null hypothesis* test on the parameter estimates. The aim is to infer whether a given parameter estimate *significantly* departs, or not, with respect to a given error probability, from a given value (which is most often 0, but not always). This inferential procedure is usually implemented by using the so-called *Wald Chi-squared test*, presented by [GRE 96]. This test is based on the one degree of freedom chi-squared distribution followed by a Gaussian r.v. of null expectation and variance unity.

### 5.2.3. *The Yule–Weibull–Cox intensity*

For practical applications, the *Yule–Weibull–Cox* intensity related to a technical object at age $t$, characterized by covariates gathered in the vector $\mathbf{Z}$, and that has undergone $j$ failures, is defined as the product of three factors:

– Yule factor $1 + \alpha j$ that ensures the dependence on the number $j$ of past events;

– Weibull factor $\delta t^{\delta-1}$ accounting for ageing, modeled as a power function of time;

– Cox factor $e^{\mathbf{Z}^T\beta}$ accounting for covariate effects.

The Yule–Weibull–Cox intensity (with $q$ covariates) is defined by the formula:

$$\mathrm{E}_\theta\left(\mathrm{d}N(t) \mid N(t-) = j, \mathbf{Z}\right) = (1 + j\alpha)\delta t^{\delta-1} e^{\mathbf{Z}^T\beta} \mathrm{d}t \qquad [5.4]$$

with:

$$\boldsymbol{\theta} = \left(\alpha, \delta, \boldsymbol{\beta}^{\mathrm{T}}\right)^{\mathrm{T}}$$

$$\alpha > 0,\ \delta \geqslant 1,\ \mathbf{Z} = (1\ Z_1\ Z_2 \ldots Z_q)^{\mathrm{T}},\ \boldsymbol{\beta} = (\beta_0\ \beta_1\ \beta_2 \ldots \beta_q)^{\mathrm{T}} \in \mathbb{R}^{q+1}$$

The "1" as first component of $\mathbf{Z}$ and the corresponding $\beta_0$ in the vector of regression coefficients $\boldsymbol{\beta}$ specifies the baseline intensity $(1 + \alpha j)\delta t^{\delta-1}e^{\beta_0}$.

### 5.2.4. *Null hypothesis test implemented for the Yule–Weibull–Cox intensity*

In the case of an unconstrained parameter $\theta_k$, the test statistic is calculated as the ratio of the squared difference between the estimate and the test value, to the estimation variance:

$$\frac{\left(\hat{\theta}_k - \theta_{k0}\right)^2}{\widehat{\mathrm{Var}}(\hat{\theta}_k)} \sim \chi_1^2$$

where $\widehat{\mathrm{Var}}(\hat{\theta}_k)$ is the $k^{\mathrm{th}}$ diagonal term of the covariance matrix $\widehat{\mathrm{Var}}(\hat{\boldsymbol{\theta}})$ defined by equation [5.3]. This method concerns the $\beta$ values associated with the above LEYP intensity covariates. It is relevant to test the $\hat{\beta}$ estimates against the test value 0, which means the lack of effect of the covariate.

In the case of constrained parameters, as $\alpha > 0$ and $\delta \geqslant 1$, the asymptotical normality assumption may not hold, especially if the estimation variance is high. It is then preferable to use the so-called *likelihood ratio test*. Concerning $\alpha$, this involves estimating with the same dataset the alternative model with $\alpha = 0$, i.e. the NHPP, calculate the $LR$ statistic, which is twice the difference of the log-likelihoods:

$$LR = 2\left(\ln L_{\mathrm{LEYP}} - \ln L_{\mathrm{NHPP}}\right)$$

and assume that, under the null hypothesis $\alpha = 0$, the $LR$ statistic is chi-squared distributed, with a degree of freedom (DF) equal to the difference between the number of parameters of the alternative models LEYP and NHPP, hence 1 DF. The same test procedure can be used for $\delta$, by considering the null hypothesis model $\delta = 1$, i.e. no ageing.

### 5.2.5. *Parameter estimation algorithm*

To find the optimal parameter values that maximize the likelihood function (the *objective function* of the optimization process), the so-called *Nelder–Mead algorithm*, initially proposed by [NEL 65] and described by [PRE 02], has empirically proven to work well for LEYP parameter estimation. For a model with $p$ parameters, this algorithm explores the space of parameter possible values using a polyhedron (also called sometimes *simplex*) with $p + 1$ vertices, that moves sequentially by undergoing basic geometric transforms: reflection of one vertex with respect to the hyperplan of the $p$ remaining vertices, expansion or contraction; the optimization process eventually contracts the polyhedron around the optimum, and stops when the maximum difference of the objective function values among the vertices is below a given threshold. The constraints $\alpha > 0$ and $\delta \geqslant 1$ are taken into account by using a *box-constrained* version of the Nelder–Mead algorithm. It consists of simply replacing after each transform of the polyhedron the $\alpha$ values that may have become negative on some vertices by a small positive value ($10^{-4}$ in practice), and by 1 the $\delta$ values that may have fallen below 1.

The Nelder–Mead algorithm has the benefit of necessitating no calculation of derivatives of the objective function with respect to the parameters. The Hessian (second order derivatives of the objective function) is nonetheless numerically evaluated at the optimum point in order to provide an estimate of the estimate covariance matrix. Extensive random experiments with synthetic datasets generated with known parameters never suggested possible convergence problems.

## 5.3. Validation of the estimation procedure

Estimating LEYP parameters from an either actual or synthetic dataset involves the implementation of a computer code, the complexity of which may be a source of potential errors, both numerical or algorithmic. In order to ensure the validity of the code, it is advisable to generate a synthetic dataset from known LEYP parameters, then to carry out the coded estimation procedure, and lastly check that the obtained estimates are reasonably close to the theoretical estimates. As this code validation method involves the ability to generate a synthetic LEYP compliant dataset, we will provide a formula for the theoretical distribution of inter-event times.

### 5.3.1. *Conditional distribution of the inter-event time*

The r.v. $T_j$ is the time at which the $j^{\text{th}}$ failure occurs . The inter-event time is the random time $X_{j+1} = T_{j+1} - T_j$ elapsed between $j^{\text{th}}$ failure and the $(j+1)^{\text{th}}$ one, as illustrated by Figure 5.1.

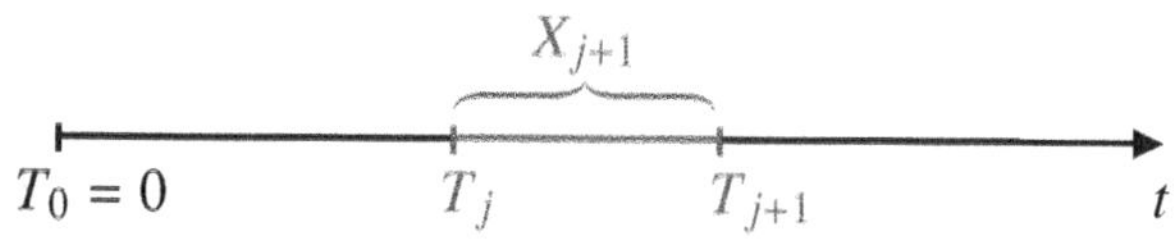


**Figure 5.1.** *Inter-failure time*

The following proposition characterizes the conditional distribution of $X_{j+1}$.

PROPOSITION 5.3.– The conditional survival function of $X_{j+1}$ given the $j^{\text{th}}$ failure time is.

$$P\left(X_{j+1} > x \mid T_j = t_j\right) = \exp\left(-(1+\alpha j)\left[\Lambda(t_j + x) - \Lambda(t_j)\right]\right)$$

PROOF.–

The sought conditional survival is the conditional probability of no failure within $]t_j, t_j + x]$ given $N(t_j) = j$, which is calculated using proposition 4.1:

$$\begin{aligned} &P\left(X_{j+1} > x \mid T_j = t_j\right) \\ &= P\left(N(t_j + x) - N(t_j) = 0 \mid N(t_j) = j\right) \\ &= \exp\left(-(1+\alpha j)\left[\Lambda(t_j + x) - \Lambda(t_j)\right]\right) \end{aligned}$$

□

### 5.3.2. *LEYP event simulation*

A sequence of events which occur according to a LEYP with known parameters can be simulated from a pseudo-random sequence of numbers uniformly distributed within the interval $[0, 1]$. The classical *inverse transform* method, presented by [ROS 97], can be used to that end, which only requires the knowledge of the analytical form of the inverse of the cumulated probability function of $X_j$.

Let $U$ be a r.v. uniformly distributed on $[0, 1]$, and $X$ a real r.v. with cumulative distribution function $F(x) = \mathrm{P}(X \leqslant x)$. As $F()$ is monotonically increasing, and as:

$$U \sim \mathcal{U}_{[0,1]} \Rightarrow \mathrm{P}(U \leqslant u) = u,$$

$F^{-1}(U)$ follows then the same distribution as $X$:

$$\mathrm{P}\left(F^{-1}(U) \leqslant x\right) = \mathrm{P}\left(F(F^{-1}(U)) \leqslant F(x)\right) = \mathrm{P}(U \leqslant F(x)) = F(x).$$

The following proposition makes it possible to simulate recursively a sequence of realizations of random failure times $T_j$:

PROPOSITION 5.4.– Let $(u_j : j \in \mathbb{N}^*)$ be a sequence of independent uniformly distributed realizations of $\mathcal{U}_{[0,1]}$, and $t_j$ a realization of $T_j$, then:

$$t_{j+1} = \Lambda^{-1}\left(\Lambda(t_j) - \frac{\ln(1 - u_{j+1})}{1 + \alpha j}\right)$$

is a realization of $T_{j+1}$.

PROOF.–

We apply proposition 5.3:

$$1 - \exp\left(-(1 + \alpha j)\left[\Lambda(t_{j+1}) - \Lambda(t_j)\right]\right) = u_{j+1}$$

$$\Rightarrow \Lambda(t_{j+1}) - \Lambda(t_j) = -\frac{\ln(1 - u_{j+1})}{1 + \alpha j}$$

$$\Rightarrow t_{j+1} = \Lambda^{-1}\left(\Lambda(t_j) - \frac{\ln(1 - u_{j+1})}{1 + \alpha j}\right)$$

□

## 5.4. LEYP model goodness of fit

Usual methods for assessing the model goodness-of-fit are based on:

– regression residuals in the case of general regression models (linear or nonlinear);

– deviance residuals in the case of generalized regression models (e.g. count data analysis based on Poisson distribution – see [MCC 89]);

– martingale residuals in the case of survival data analysis models (see [THE 90]).

Martingale residuals method have been extended to counting process models [LAW 95, AND 93, AAL 08], and defined for this purpose as the difference:

$$N(t) - \int_0^t \mathrm{E}\left(\mathrm{d}N(u) \mid N(u-)\right).$$

Unfortunately, when observation starts at age $a > 0$, such residuals cannot be calculated since $N(t)$ is unknown (only $N(t) - N(a)$ is observed). It is useful yet to compute for each segment the residual:

$$N(b) - N(a) - \int_a^b \mathrm{E}\left(\mathrm{d}N(u) \mid N(u-)\right),$$

and check the segment data related to the highest values, as they may allow us to detect absurd data.

The model goodness-of-fit can be globally assessed by graphically comparing the empirical and theoretical failure rates averaged over the sample of objects, and plotted against age, as explained in [AND 93].

In the water network IAM context, the technical objects considered are network segments, also equivalently named pipes, defined as contiguous pipeline elements, homogeneous with respect to their age, material and pipe diameter. An $n$-sample of segments indexed by $i$ is considered, with length denoted by $l_i$. Each segment is characterized by the covariate vector $\boldsymbol{Z}_i$, and has been observed within $[a_i, b_i]$ where its counting process $N_i(t)$ has recorded $m_i$ failures at times $\left(t_{ij}\right)_{j \in \{1,\ldots,m_i\}}$.

The empirical estimator $\hat{\rho}(t)$ of the average failure rate at age $t$ is calculated as the weighted number of failures observed to occur within

$[t - \Delta t, t + \Delta t]$ divided by the total length of the segments observed within the same age interval, and by the bandwidth $\Delta t$:

$$\hat{\rho}(t) = \left( \sum_{i=1}^{n} \sum_{j=1}^{m_i} K\left(\frac{t - t_{ij}}{\Delta t}\right) \right) \left( \sum_{i=1}^{n} \mathrm{I}(a_i \leq t \leq b_i)\, l_i \right)^{-1} (\Delta t)^{-1} \qquad [5.5]$$

$\hat{\rho}(t)$ is a Nelson–Aalen empirical estimator of the increments of $N(t)$, smoothed by the Epanechnikov kernel $K(x)$ with bandwidth $\Delta t$:

$$K(x) = 0.75(1 - x^2)\mathrm{I}(|\, x \,| \leq 1), x \in \mathbb{R}$$

Smoothing is especially useful at older ages, poorly represented in the available data, which may generate chaotic inter-annual variations. Other smoothing methods could have been used (e.g. uniform or biweight kernels), but the Epanechnikov kernel is a good trade-off that uses adjacent ages information but gives a more prominent weight to the current one. It must also be noticed that the empirical failure rate defined by equation [5.5] complies with the practical concept of failure rate used by water utilities and defined as a number of failures by time unit and by network segment length unit.

To compute a confidence interval for $\hat{\rho}(t)$, its variance can be estimated as:

$$\widehat{\mathrm{Var}}\hat{\rho}(t) = \left( \sum_{i=1}^{n} \sum_{j=1}^{m_i} K^2\left(\frac{t - t_{ij}}{\Delta t}\right) \right) \left( \sum_{i=1}^{n} \mathrm{I}(a_i \leq t \leq b_i)\, l_i \right)^{-2} (\Delta t)^{-2}$$

The theoretical average conditional failure rate at age $t$ is calculated as (see proposition 5.1 for the conditional intensity):

$$\sum_{i=1}^{n} \mathrm{E}_\theta \left(\mathrm{d}N_i(t) \mid N_i(t-) - N_i(a_i), \mathbf{Z}_i\right) \mathrm{I}(a_i \leq t \leq b_i) \left( \sum_{i=1}^{n} \mathrm{I}(a_i \leq t \leq b_i)\, l_i \right)^{-1} \qquad [5.6]$$

## 5.5. Validating LEYP model predictions

A method to validate the model predictions has been proposed by [LEG 02] (see also [REN 12] and [LEG 14]); it consists of:

– calibrating the model using, for example, the first 80% of the duration of the available observation window;

– then predicting for each pipe the number of failures in the last 20% of the window.

This makes it possible to compare in the last 20% of the observation window the actual number of failures versus the predicted number of failures. The relative comparison allows us to make an assessment of the ability of the model to detect the pipes with highest failure risks, by measuring how actual failures are concentrated on the pipes with the highest expected failure rate. This is done by building a predictive performance curve presented in the next section.

### 5.5.1. *Lorenz curve*

The building of predictive performance curves follows an outline similar to the method proposed by [LOR 05] to graphically assess the inequity of economic wealth concentration within a population. Such a curve is often called a *Lorenz curve*, or also *Lift curve* in marketing applications.

The following presentation is adapted to applications in the water network IAM context, where the technical object considered is a network segment indexed by $i$, of length $l_i$, and assumed to have been observed both:

– within calibration interval $[a_i, b_i]$, where it experienced $m_i$ failures;

– and within validation interval $[b_i, c_i]$, where it experienced $k_i$ failures.

The key point is the ranking of the pipes by decreasing expected failure rate:

$$\hat{k}_i / l_i / (c_i - b_i)$$

where:

$$\hat{k}_i = \mathrm{E}_\theta \left( N_i(c_i) - N_i(b_i) \mid N_i(b_i) - N_i(a_i) = m_i, \mathbf{Z}_i \right).$$

Considering the subset of size $i$ of pipes with highest expected failure rates, the relative risk rank weighted by the pipe length is calculated as:

$$r_{(i)} = \sum_{j=(1)}^{(i)} l_j / \sum_{j=(1)}^{(n)} l_j$$

and the relative number of failures observed within the validation interval:

$$\kappa_{(i)} = \sum_{j=(1)}^{(i)} k_j / \sum_{j=(1)}^{(n)} k_j$$

where parenthesized indices $(1), \ldots, (i), \ldots, (n)$ means that pipes are arranged in decreasing order of expected failure rate. The predictive performance curve is the graph $(r_{(i)}, \kappa_{(i)})$, as illustrated by Figure 5.2. The area $\mathscr{A}$ under the step curve is calculated as:

$$\mathscr{A} = \frac{\sum_{i=(1)}^{(n)} l_i \kappa_i}{\sum_{i=(1)}^{(n)} l_i}$$

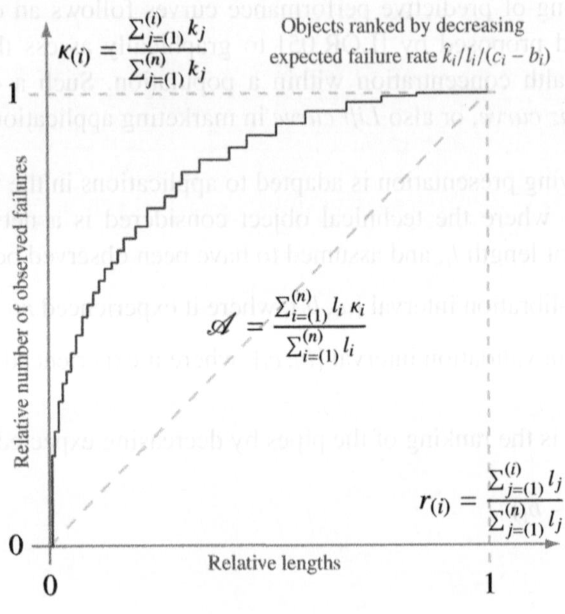


**Figure 5.2.** *Lorenz curve*

The predictive performance is all better since $\mathscr{A}$ is closer to 1. $\kappa_{(i)}$ can be interpreted as the rate of failures that would have been avoided if a corresponding rate $r_{(i)}$ of the total pipe length had been replaced, provided the replaced pipes have been chosen in priority according to their theoretical failure rate. Despite a formal resemblance with the so-called *receiver operating curve*, used by [DEB 10] for validating pipe failure predictions, the predictive performance curve cannot be similarly interpreted, since the response variable is not a Bernoulli random variable.

### 5.5.2. *Prediction bias checking*

For a sample of objects sufficiently large, for example $n \geqslant 30$, the lack of bias in the predictions can be assessed by checking that the sum of the actual numbers of failures $\sum_{i=1}^{n} k_i$ lies within 95% confidence interval of the sum of the expected numbers of failures $\sum_{i=1}^{n} \hat{k}_i$. This confidence interval can be calculated using the asymptotic normality of a sum of independent negative binomial variables with variance $\sum_{i=1}^{n} \widehat{\text{Var}}(\hat{k}_i)$, provided the variances $\widehat{\text{Var}}(\hat{k}_i)$ do not vary too much.

# 6

# Selective Survival

In the previous Chapters 4 and 5, the objects under study were assumed to remain indefinitely in service, i.e. have an infinite lifetime. This hypothesis is of course simplistic with regard to the real world, but nevertheless useful in the first step of research work. In order to go one step further toward our objective to design a tool to aid infrastructure asset management (IAM) decisions, finite service lifetimes have to be considered instead. Revisiting the LEYP model in this respect turns out to be also theoretically interesting, as the end of the service life of an asset, i.e. its decommissioning, obviously stops the observation of its failure process, and is then an important cause of right-censoring. If repeated failures can motivate the decommissioning, right-censoring has then to be considered, at least partly, as informative (or dependent), which necessitates an adapted mathematical treatment.

## 6.1. Left-truncation, right-censoring and decommissioning decisions

In order to practically implement a failure model within an IAM decision process, it is necessary to understand why the elementary components of the infrastructure may have shorter or longer service lifetimes. More specifically, in the case of water networks, water mains can be decommissioned for various reasons, that can be grouped for theoretical grounds into two main categories, according to whether the decommissioning is motivated by the performance degradation, or not:

– we call *selective*, the decommissioning which is motivated by repeated failures of a given object; this kind of decommissioning results indeed from a *decision* made because the observed failure rate of the considered object exceeds a threshold, which is a function of potential impacts of failures on users and environment;

– the decommissioning is called *constrained* when it is independent of the object condition; this kind of decommissioning is most often dictated by land management decisions, which involve the considered object to be moved or reconfigurated.

It is thought that there exists in practice an intermediate case where a decommissioning decision is made on the occasion of land management work with little regard for the suspected object condition, but as a precaution, with the aim of taking the opportunity of third party works and potential cost reduction. Such *opportunity* decommissioning most often concerns objects of a certain age, and can be considered as an age-dependent constrained decommissioning.

Selective decommissioning involves a selection phenomenon of the technical objects on their robustness, which will be called *selective survival*. This makes it tricky to estimate LEYP model parameter, and all the more since actual failure data are left-truncated, as already mentioned in sections 1.6 and 4.2. Moreover, in the case where the observation window is terminated because of the decommissioning, the right-censoring of the observation of the failure process cannot be considered as *non-informative* or *independent*. As a major consequence of left-truncation and informative right-censoring, both dependent on the failure process itself, the objects with high failure rate are likely to be under-represented in actual datasets. It is therefore of utmost importance to build an enhanced version of the LEYP model that accounts for both processes of failure and decommissioning.

## 6.2. Coupling failure and decommissioning processes: LEYP2s model

In order to couple the failure and decommissioning processes, the counting process $R(t)$ devoted to decommissioning events is introduced in addition to the failure counting process $N(t)$. The coupling between both processes is ensured by making the decommissioning intensity linearly depend on $N(t-)$. This gives rise to the so-called *LEYP2s* model, enhanced version of LEYP that accounts for selective survival.

DEFINITION 6.1.– *The LEYP2s model is defined by coupling the failure and decommissioning intensity functions as follows:*

$$\forall t \in \mathbb{R}_+,$$
$$\begin{cases} N(0) = 0 \\ \mathrm{E}\,(\mathrm{d}N(t) \mid N(t-)) = (1 + \alpha N(t-))\,\lambda(t)\mathrm{d}t \\ R(0) = 0 \\ \mathrm{E}\,(\mathrm{d}R(t) \mid N(t-)) = (\psi(t) + \phi N(t-))\,\mathrm{d}t \end{cases}$$
*with:* $\alpha > 0, \quad \lambda(t) \geqslant 0, \quad \psi(t) \geqslant 0, \quad \phi > 0$

The parameter function $\psi(t)$ and scalar $\phi$ relate, respectively, to the constrained and selective decommissionings. For the constrained decommissioning, a function is considered instead of a scalar to account also for opportunity decommissioning.

The decommissioning intensity is hence designed as a *competing risk model*, in which two decommissioning causes compete independently of each other (see [AAL 08] p. 18). Two competing decommissioning processes can then be defined; they will be denoted, respectively, $R_C(t)$ and $R_S(t)$, for the constrained and selective decommissionings, the intensities of which are:

$$\mathrm{E}\,(\mathrm{d}R_C(t)) = \psi(t)\mathrm{d}t$$

$$\mathrm{E}\,(\mathrm{d}R_S(t) \mid N(t-)) = \phi N(t-)\mathrm{d}t$$

The following additivity of intensities characterises the competing risk concept:

$$\mathrm{E}\,(\mathrm{d}R(t) \mid N(t-)) = \mathrm{E}\,(\mathrm{d}R_C(t)) + \mathrm{E}\,(\mathrm{d}R_S(t) \mid N(t-))$$

## 6.3. LEYP2s discretization scheme

In order to obtain a tractable tool for practical applications, we need to derive the distributional properties of the LEYP2s model for a given technical object conditionally on its survival until the beginning of the observation window $[a, b]$, i.e. given $R(a-) = 0$. However, despite the apparent simplicity of the LEYP2s in definition 6.1, the study of this process is difficult to carry out in continuous time.

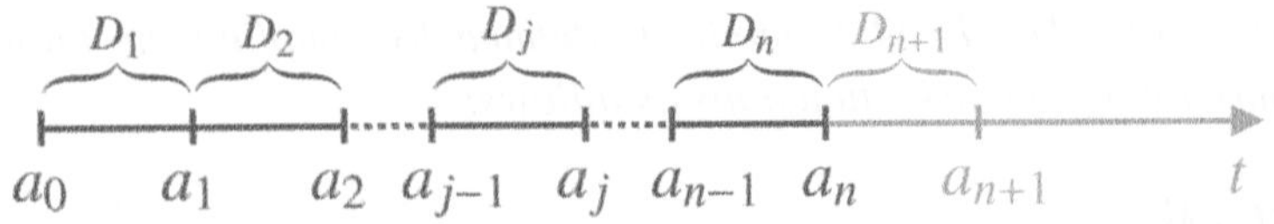


**Figure 6.1.** *LEYP2s discretization scheme on interval* $[0, a[$

We have then chosen to partition the interval $[0, a[$ according to Figure 6.1, with by convention $a_0 = 0$ and $a_n = a$. For each subinterval $[a_{j-1}, a_j[$, the following notations are adopted:

$$D_j = N(a_j-) - N(a_{j-1})$$

$$\Delta\mu_j = \mu(a_j) - \mu(a_{j-1})$$

$$\psi_j = e^{-\int_{a_{j-1}}^{a_j} \psi(t)dt} \quad [6.1]$$

$$\phi_j = e^{-(a_j - a_{j-1})\phi} \quad [6.2]$$

It is assumed that:

– decommissioning can only occur at times $a_j$;

– the probability of no decommissioning at $a_j$ given none occured before and given the numbers of failures within intervals $[a_{i-1}, a_i[$, for $i = 1, \ldots, j$, is:

$$P\left(R(a_j) - R(a_{j-1}) = 0 \mid \bigcap_{i=1}^{j} D_i = d_i\right) = \psi_j \phi_j^{\sum_{i=1}^{j} d_i}$$

PROPOSITION 6.1.– The discretization scheme tends to a LEYP2s when $n \to \infty$ while $\max_j \left(a_j - a_{j-1}\right) \to 0+$.

PROOF.–

Setting indeed $a_j = t$ in the above, noting that $\sum_{i=1}^{j} D_j$ is then $N(t-)$, and as:

$$\lim_{x \to 0+} 1 - \exp^{-x} = x,$$

allows us to check that:

$$\lim_{a_j - a_{j-1} \to 0+} 1 - \psi_j \phi_j^{\sum_{i=1}^{j} D_j} = (\psi(t) + \phi N(t-))\, \mathrm{d}t$$

$$= \mathrm{E}\,(\mathrm{d}R(t) \mid N(t-)) \qquad \square$$

## 6.4. Failure and decommissioning probabilities

Whereas definition 6.1 makes LEYP2s Markovian, the probability $\mathrm{P}\,(R(t) = 0)$ that the technical object has not been decommissioned yet at age $t$ depends on the specific trajectory followed by the failure process $N(s), s < t$. We will see, however, that the discretization scheme makes it nevertheless possible to establish simple probabilistic results for both $R(t)$ and $N(t)$ processes.

### 6.4.1. *Probability of no decommissioning*

Within the discretization framework illustrated by Figure 6.1 we will establish first the following marginal probability that no decommissioning occurs up to a given time.

PROPOSITION 6.2.– The marginal probability that no decommissioning occurs up to time $a_n$ is:

$$\mathrm{P}\,(R(a_n) = 0) = \left(\prod_{j=1}^{n} \psi_j\right)\left(\mu_n - \sum_{j=1}^{n}\left(\prod_{i=j}^{n} \phi_i\right)\Delta\mu_j\right)^{-\alpha^{-1}}$$

PROOF.– A specific discretized trajectory of the failure process between $a_0$ and $a_n$ is completely defined by an $n$-tuple $(d_1, \ldots, d_n)$, and the corresponding conditional probability is hence:

$$\mathrm{P}\left(R(a_n) = 0 \mid \bigcap_{j=1}^{n} D_j = d_j\right) = \prod_{j=1}^{n} \psi_j \phi_j^{\sum_{i=1}^{j} d_i}$$

which can be rewritten:

$$= \prod_{j=1}^{n} \psi_j \left(\prod_{i=j}^{n} \phi_i\right)^{d_j}$$

Using law of total probability:

$$
\begin{aligned}
&\mathrm{P}\left(R(a_n)=0\right) \\
&= \sum_{d_1=0}^{\infty}\cdots\sum_{d_n=0}^{\infty} \mathrm{P}\left(R(a_n)=0 \mid \bigcap_{j=1}^{n} D_j = d_j\right)\mathrm{P}\left(\bigcap_{j=1}^{n} D_j = d_j\right)
\end{aligned}
$$

Then, from proposition 4.9, with $a_0 = 0$:

$$
\begin{aligned}
&= \sum_{d_1=0}^{\infty}\cdots\sum_{d_n=0}^{\infty}\prod_{j=1}^{n}\psi_j\left(\prod_{i=j}^{n}\phi_i\right)^{d_j}\frac{\Gamma(\alpha^{-1}+\sum_{j=1}^{n} d_j)}{\Gamma(\alpha^{-1})\prod_{j=1}^{n} d_j!}\left(\frac{1}{\mu_n}\right)^{\alpha^{-1}}\prod_{j=1}^{n}\left(\frac{\Delta\mu_j}{\mu_n}\right)^{d_j} \\
&= \left(\prod_{j=1}^{n}\psi_j\right)\left(\frac{1}{\mu_n}\right)^{\alpha^{-1}}\sum_{d_1=0}^{\infty}\cdots\sum_{d_n=0}^{\infty}\frac{\Gamma(\alpha^{-1}+\sum_{j=1}^{n} d_j)}{\Gamma(\alpha^{-1})\prod_{j=1}^{n} d_j!}\prod_{j=1}^{n}\left(\frac{\left(\prod_{i=j}^{n}\phi_i\right)\Delta\mu_j}{\mu_n}\right)^{d_j}
\end{aligned}
$$

and using equation [2.8]:

$$
\begin{aligned}
&= \left(\prod_{j=1}^{n}\psi_j\right)\left(\frac{1}{\mu_n}\right)^{\alpha^{-1}}\left(1-\sum_{j=1}^{n}\frac{\left(\prod_{i=j}^{n}\phi_i\right)\Delta\mu_j}{\mu_n}\right)^{-\alpha^{-1}} \\
&= \left(\prod_{j=1}^{n}\psi_j\right)\left(\mu_n-\sum_{j=1}^{n}\left(\prod_{i=j}^{n}\phi_i\right)\Delta\mu_j\right)^{-\alpha^{-1}}
\end{aligned}
$$

□

Passing to the limit ($n \to \infty$ while $\max_{j=1,\ldots,n}\left(a_j - a_{j-1}\right) \to 0+$), and setting in the above:

$$
a_0 = 0, \quad a_n = a,
$$

the following proposition related to LEYP2s can be infered.

PROPOSITION 6.3.– The marginal probability that no decommissioning occurs up to time $a-$ is:

$$
\mathrm{P}\left(R(a-)=0\right) = \xi(a)\left(\mu(a)-\nu(a)\right)^{-\alpha^{-1}}
$$

where:

$$\xi(a) = e^{-\int_0^a \psi(t)dt} \tag{6.3}$$

and:

$$\nu(a) = \int_0^a e^{-(a-t)\phi} d\mu(t) \tag{6.4}$$

PROOF.–

The proposition is a straightforward consequence of proposition 6.2 and notations defined by equations [6.1] and [6.2] □

### 6.4.2. *Distribution of* $N(b) - N(a)$ *given* $R(a-) = 0$

We will now establish the conditional counterpart of proposition 4.3 given $R(a-)$, i.e. calculate the conditional distribution of the number of failures within an interval given no decommissioning occured up to the beginning of this interval. We first consider the discretized problem (see Figure 6.1).

PROPOSITION 6.4.– The conditional distribution of $D_{n+1}$ given $R(a_n) = 0$ is negative binomial:

$$D_{n+1} \mid R(a_n) = 0 \sim \mathcal{NB}\left(\alpha^{-1}, \frac{\mu_n - \sum_{j=1}^{n}\left(\prod_{i=j}^{n}\phi_i\right)\Delta\mu_j}{\mu_{n+1} - \sum_{j=1}^{n}\left(\prod_{i=j}^{n}\phi_i\right)\Delta\mu_j}\right)$$

PROOF.–

Using Bayes' theorem:

$$\begin{aligned} &P(D_{n+1} = d_{n+1} \mid R(a_n) = 0) \\ &= \frac{P(R(a_n) = 0 \mid D_{n+1} = d_{n+1})P(D_{n+1} = d_{n+1})}{P(R(a_n) = 0)} \end{aligned}$$

The numerator can be calculated using law of total probability:

$$P(R(a_n) = 0 \mid D_{n+1} = d_{n+1})P(D_{n+1} = d_{n+1})$$

$$= \sum_{d_1=0}^{\infty} \cdots \sum_{d_n=0}^{\infty} \mathrm{P}\left(R(a_n) = 0 \mid \bigcap_{j=1}^{n} D_j = d_j, D_{n+1} = d_{n+1}\right)$$

$$\times \mathrm{P}\left(\bigcap_{j=1}^{n} D_j = d_j \mid D_{n+1} = d_{n+1}\right) \mathrm{P}\left(D_{n+1} = d_{n+1}\right)$$

$$= \sum_{d_1=0}^{\infty} \cdots \sum_{d_n=0}^{\infty} \prod_{j=1}^{n} \psi_j \left(\prod_{i=j}^{n} \phi_i\right)^{d_j} \frac{\mathrm{P}\left(\bigcap_{j=1}^{n} D_j = d_j, D_{n+1} = d_{n+1}\right)}{\mathrm{P}\left(D_{n+1} = d_{n+1}\right)} \mathrm{P}\left(D_{n+1} = d_{n+1}\right)$$

Then, from proposition 4.9 at order $n+1$, with $a_0 = 0$:

$$= \sum_{d_1=0}^{\infty} \cdots \sum_{d_n=0}^{\infty} \prod_{j=1}^{n} \psi_j \left(\prod_{i=j}^{n} \phi_i\right)^{d_j} \frac{\Gamma(\alpha^{-1} + \sum_{j=1}^{n+1} d_j)}{\Gamma(\alpha^{-1}) \prod_{j=1}^{n+1} d_j!} \left(\frac{1}{\mu_{n+1}}\right)^{\alpha^{-1}} \prod_{j=1}^{n+1} \left(\frac{\Delta\mu_j}{\mu_{n+1}}\right)^{d_j}$$

$$= \left(\prod_{j=1}^{n} \psi_j\right) \frac{\Gamma(\alpha^{-1} + d_{n+1})}{\Gamma(\alpha^{-1}) d_{n+1}!} \left(\frac{1}{\mu_{n+1}}\right)^{\alpha^{-1}} \left(\frac{\Delta\mu_{n+1}}{\mu_{n+1}}\right)^{d_{n+1}}$$

$$\times \sum_{d_1=0}^{\infty} \cdots \sum_{d_n=0}^{\infty} \frac{\Gamma(\alpha^{-1} + d_{n+1} + \sum_{j=1}^{n} d_j)}{\Gamma(\alpha^{-1} + d_{n+1}) \prod_{j=1}^{n} d_j!} \prod_{j=1}^{n} \left[\left(\prod_{i=j}^{n} \phi_i\right) \left(\frac{\Delta\mu_j}{\mu_{n+1}}\right)\right]^{d_j}$$

Then, using equation [2.8]:

$$= \left(\prod_{j=1}^{n} \psi_j\right) \frac{\Gamma(\alpha^{-1} + d_{n+1})}{\Gamma(\alpha^{-1}) d_{n+1}!} \left(\frac{1}{\mu_{n+1}}\right)^{\alpha^{-1}} \left(\frac{\Delta\mu_{n+1}}{\mu_{n+1}}\right)^{d_{n+1}}$$

$$\left(1 - \sum_{j=1}^{n} \left(\prod_{i=j}^{n} \phi_i\right) \frac{\Delta\mu_j}{\mu_{n+1}}\right)^{-(\alpha^{-1} + d_{n+1})}$$

$$= \left(\prod_{j=1}^{n} \psi_j\right) \frac{\Gamma(\alpha^{-1} + d_{n+1})}{\Gamma(\alpha^{-1}) d_{n+1}!} \frac{(\Delta\mu_{n+1})^{d_{n+1}}}{\left(\mu_{n+1} - \sum_{j=1}^{n} \left(\prod_{i=j}^{n} \phi_i\right) \Delta\mu_j\right)^{\alpha^{-1} + d_{n+1}}}$$

And finally, dividing by denominator (calculated by proposition 6.3) yields:

$$\mathrm{P}\left(D_{n+1} = d_{n+1} \mid R(a_n) = 0\right)$$

$$= \frac{\Gamma(\alpha^{-1} + d_{n+1})}{\Gamma(\alpha^{-1}) d_{n+1}!} \left( \frac{\mu_n - \sum_{j=1}^{n} \left( \prod_{i=j}^{n} \phi_i \right) \Delta\mu_j}{\mu_{n+1} - \sum_{j=1}^{n} \left( \prod_{i=j}^{n} \phi_i \right) \Delta\mu_j} \right)^{\alpha^{-1}}$$

$$\left( \frac{\Delta\mu_{n+1}}{\mu_{n+1} - \sum_{j=1}^{n} \left( \prod_{i=j}^{n} \phi_i \right) \Delta\mu_j} \right)^{d_{n+1}}$$

□

Passing to the limit ($n \to \infty$ while $\max_{j=1,\dots,n} \left( a_j - a_{j-1} \right) \to 0+$), and setting in the above:

$$a_0 = 0, \quad a_n = a, \quad a_{n+1} = b,$$

yields the following proposition.

PROPOSITION 6.5.– The conditional distribution of $N(b) - N(a)$ given $R(a-) = 0$ is negative binomial:

$$N(b) - N(a) \mid R(a-) = 0 \sim \mathcal{NB}\left( \alpha^{-1}, \frac{\mu(a) - \nu(a)}{\mu(b) - \nu(a)} \right)$$

PROOF.–

The proposition is a straightforward consequence of proposition 6.4. □

REMARK 6.1.– Setting $\phi = 0$ in proposition 6.5 yields proposition 4.3, as $\int_0^a \mathrm{e}^{-(a-t)\phi} \mathrm{d}\mu(t)$ becomes then $\mu(a) - 1$. △

### 6.4.3. ***Conditional probability of*** $R(a-) = 0$ ***given*** $N(b) - N(a)$

The converse of previous proposition 6.5 is also useful.

PROPOSITION 6.6.– The conditional probability of $R(a-) = 0$ given $N(b) - N(a)$ is:

$$\mathrm{P}\left( R(a-) = 0 \mid N(b) - N(a) = m \right) = \xi(a) \left( \frac{\mu(b) - \mu(a) + 1}{\mu(b) - \nu(a)} \right)^{\alpha^{-1} + m}$$

PROOF.–

Using Bayes' theorem:

$$\mathrm{P}(R(a-)) = 0 \mid N(b) - N(a) = m$$
$$= \frac{\mathrm{P}(N(b) - N(a) = m \mid R(a-) = 0)\,\mathrm{P}(R(a-) = 0)}{\mathrm{P}(N(b) - N(a) = m)}$$

Applying propositions 6.5, 6.3 and 4.3:

$$= \frac{\Gamma(\alpha^{-1} + m)}{\Gamma(\alpha^{-1})m!} \left(\frac{\mu(a) - \nu(a)}{\mu(b) - \nu(a)}\right)^{\alpha^{-1}} \left(\frac{\mu(b) - \mu(a)}{\mu(b) - \nu(a)}\right)^{m} \xi(a)\,(\mu(a) - \nu(a))^{\alpha^{-1}} \times$$
$$\times \left(\frac{\Gamma(\alpha^{-1} + m)}{\Gamma(\alpha^{-1})m!} \left(\frac{1}{\mu(b) - \mu(a) + 1}\right)^{\alpha^{-1}} \left(\frac{\mu(b) - \mu(a)}{\mu(b) - \mu(a) + 1}\right)^{m}\right)^{-1}$$
$$= \xi(a) \left(\frac{\mu(b) - \mu(a) + 1}{\mu(b) - \nu(a)}\right)^{\alpha^{-1}+m} \qquad \square$$

REMARK 6.2.– In proposition 6.6, the conditional probability of undergoing no decommissioning within interval $[0, a[$ given the number of failures within interval $[a, b]$ does not depend on the trajectory of the failure process within that interval. As a consequence, if we partition $[a, b[$[1] into $n$ ajacent non-overlapping subintervals:

$$[a, b[ = \bigcup_{j=1}^{n} [b_{j-1}, b_j[$$

where $b_0 = a$ and $b_n = b$, we can then write as:

$$\mathrm{P}\left(R(a-) = 0 \mid \bigcap_{j=1}^{n} N(b_j) - N(b_{j-1})\right) = \mathrm{P}(R(a-) = 0 \mid N(b) - N(a)) \qquad \triangle$$

1 Time point $b$ is excluded for sole notation convenience, without loss of generality for our purpose.

REMARK 6.3.– A discretized version of proposition 6.6 will be needed to prove proposition 6.8 in the next section 6.4.6:

$$P\left(R(a_n = 0 \mid D_{n+1} = d_{n+1})\right) = \left(\prod_{j=1}^{n} \psi_j\right)\left(\frac{1 + \Delta\mu_{n+1}}{\mu_{n+1} - \sum_{j=1}^{n}\left(\prod_{i=j}^{n}\phi_i\right)\Delta\mu_j}\right)^{\alpha^{-1}+d_{n+1}}$$

△

We have yet to establish two propositions that will be useful in practical LEYP2s applications to carry out conditional predictions in a future time interval given the number of failures within an available observation window.

### 6.4.4. *Conditional distribution of* $N(c) - N(b)$ *given* $N(b) - N(a)$ *and* $R(a-) = 0$

We will establish thus the conditional counterpart of proposition 4.5 given $R(a-)$.

PROPOSITION 6.7.– The conditional distribution of $N(c)-N(b)$ given $N(b)-N(a)$ and $R(a-) = 0$ is negative binomial:

$$N(c) - N(b) \mid N(b) - N(a) = m, R(a-) = 0 \sim \mathcal{NB}\left(\alpha^{-1} + m, \frac{\mu(b) - \nu(a)}{\mu(c) - \nu(a)}\right)$$

PROOF.–

Using Bayes' theorem:

$$\begin{aligned} &P\left(N(c) - N(b) = k \mid N(b) - N(a) = m, R(a-) = 0\right) \\ &= P\left(R(a-) = 0 \mid N(c) - N(b) = k, N(b) - N(a) = m\right) \\ &\times \frac{P\left(N(c) - N(b) = k \mid N(b) - N(a) = m\right)}{P\left(R(a-) = 0 \mid N(b) - N(a) = m\right)} \end{aligned}$$

But, as stressed by remark 6.2:

$$\begin{aligned} &P\left(R(a-)\right) = 0 \mid N(c) - N(b) = k, N(b) - N(a) = m \\ &\qquad = P\left(R(a-) = 0 \mid N(c) - N(a) = m + k\right) \end{aligned}$$

Then:

$$\mathrm{P}(N(c) - N(b) = k \mid N(b) - N(a) = m, R(a-) = 0)$$

$$= \xi(a)\left(\frac{\mu(c) - \mu(a) + 1}{\mu(c) - \nu(a)}\right)^{\alpha^{-1}+m+k}$$

$$\times \frac{\Gamma(\alpha^{-1} + m + k)}{\Gamma(\alpha^{-1} + m)k!}\left(\frac{\mu(b) - \mu(a) + 1}{\mu(c) - \mu(a) + 1}\right)^{\alpha^{-1}+m}\left(\frac{\mu(c) - \mu(b)}{\mu(c) - \mu(a) + 1}\right)^{k}$$

$$\times \left(\xi(a)\left(\frac{\mu(b) - \mu(a) + 1}{\mu(b) - \nu(a)}\right)^{\alpha^{-1}+m}\right)^{-1}$$

$$= \frac{\Gamma(\alpha^{-1} + m + k)}{\Gamma(\alpha^{-1} + m)k!}\left(\frac{\mu(b) - \nu(a)}{\mu(c) - \nu(a)}\right)^{\alpha^{-1}+m}\left(\frac{\mu(c) - \mu(b)}{\mu(c) - \nu(a)}\right)^{k} \qquad \square$$

### 6.4.5. *Conditional distribution of* $N(d) - N(c)$ *given* $N(b) - N(a)$ *and* $R(a-) = 0$

We can now establish the conditional counterpart of proposition 4.7 given $R(a-)$, which will be proven by using proposition 6.7 twice.

PROPOSITION 6.8.– The conditional distribution of $N(d)-N(c)$ given $N(b)-N(a)$ and $R(a-) = 0$ is negative binomial:

$$N(d) - N(c) \mid N(b) - N(a) = m, R(a-)$$

$$= 0 \sim \mathcal{NB}\left(\alpha^{-1} + m, \frac{\mu(b) - \nu(a)}{\mu(d) - \mu(c) + \mu(b) - \nu(a)}\right)$$

PROOF.–

Using the law of total probability:

$$\mathrm{P}(N(d) - N(c) = k \mid N(b) - N(a) = m, R(a-) = 0)$$

$$= \sum_{j=0}^{\infty} \mathrm{P}(N(d) - N(c) = k \mid N(c) - N(b) = j, N(b) - N(a) = m, R(a-) = 0)$$

$$\times \mathrm{P}(N(c) - N(b) = j \mid N(b) - N(a) = m, R(a-) = 0)$$

$$= \sum_{j=0}^{\infty} \mathrm{P}\left(N(d) - N(c) = k \mid N(c) - N(a) = m + j, R(a-) = 0\right)$$

$$\times \mathrm{P}\left(N(c) - N(b) = j \mid N(b) - N(a) = m, R(a-) = 0\right)$$

Using proposition 6.7 twice:

$$= \sum_{j=0}^{\infty} \frac{\Gamma(\alpha^{-1} + m + j + k)}{\Gamma(\alpha^{-1} + m + j)k!} \left(\frac{\mu(c) - \nu(a)}{\mu(d) - \nu(a)}\right)^{\alpha^{-1}+m+j} \left(\frac{\mu(d) - \mu(c)}{\mu(d) - \nu(a)}\right)^{k}$$

$$\times \frac{\Gamma(\alpha^{-1} + m + j)}{\Gamma(\alpha^{-1} + m)j!} \left(\frac{\mu(b) - \nu(a)}{\mu(c) - \nu(a)}\right)^{\alpha^{-1}+m} \left(\frac{\mu(c) - \mu(b)}{\mu(c) - \nu(a)}\right)^{j}$$

$$= \frac{\Gamma(\alpha^{-1} + m + k)}{\Gamma(\alpha^{-1} + m)k!} \left(\frac{\mu(b) - \nu(a)}{\mu(d) - \nu(a)}\right)^{\alpha^{-1}+m} \left(\frac{\mu(d) - \mu(c)}{\mu(d) - \nu(a)}\right)^{k}$$

$$\times \sum_{j=0}^{\infty} \frac{\Gamma(\alpha^{-1} + m + j + k)}{\Gamma(\alpha^{-1} + m + k)j!} \left(\frac{\mu(c) - \mu(b)}{\mu(d) - \nu(a)}\right)^{j}$$

And using negative binomial power series of equation [2.6] finally yields:

$$= \frac{\Gamma(\alpha^{-1} + m + k)}{\Gamma(\alpha^{-1} + m)k!} \left(\frac{\mu(b) - \nu(a)}{\mu(d) - \nu(a)}\right)^{\alpha^{-1}+m} \left(\frac{\mu(d) - \mu(c)}{\mu(d) - \nu(a)}\right)^{k} \left(1 - \frac{\mu(c) - \mu(b)}{\mu(d) - \nu(a)}\right)^{-(\alpha^{-1}+m+k)}$$

$$= \frac{\Gamma(\alpha^{-1} + m + k)}{\Gamma(\alpha^{-1} + m)k!} \left(\frac{\mu(b) - \nu(a)}{\mu(d) - \mu(c) + \mu(b) - \nu(a)}\right)^{\alpha^{-1}+m} \left(\frac{\mu(d) - \mu(c)}{\mu(d) - \mu(c) + \mu(b) - \nu(a)}\right)^{k}$$

□

### 6.4.6. *Conditional distribution of* $N(a-)$ *given* $N(b) - N(a)$ *and* $R(a-) = 0$

We conclude this chapter with the conditional counterpart of proposition 4.4 given $R(a-) = 0$, that will be useful later on when building the LEYP2s likelihood. We consider first the discretized problem.

PROPOSITION 6.9.– The conditional distribution of $\sum_{j=1}^{n} D_j$ given $D_{n+1}$ and $R(a_n) = 0$ is negative binomial:

$$\sum_{j=1}^{n} D_j \mid D_{n+1} = d_{n+1}, R(a_n) = 0 \sim \mathcal{NB}\left(\alpha^{-1} + d_{n+1}, \frac{\mu_{n+1} - \sum_{j=1}^{n}\left(\prod_{i=j}^{n}\phi_i\right)\Delta\mu_j}{\mu_{n+1}}\right)$$

PROOF.–

We first consider the conditional distribution of:

$$\bigcap_{j=1}^{n} D_j \mid D_{n+1} = d_{n+1}, R(a_n) = 0$$

and use Bayes theorem to calculate:

$$\mathrm{P}\left(\bigcap_{j=1}^{n} D_j = d_j \mid D_{n+1} = d_{n+1}, R(a_n) = 0\right)$$
$$= \frac{\mathrm{P}\left(R(a_n) = 0 \mid \bigcap_{j=1}^{n+1} D_j = d_j\right)\mathrm{P}\left(\bigcap_{j=1}^{n} D_j = d_j \mid D_{n+1} = d_{n+1}\right)}{\mathrm{P}\left(R(a_n) = 0 \mid D_{n+1} = d_{n+1}\right)}$$

As $R(a_n)$ is completely determined by $\bigcap_{j=1}^{n} D_j = d_j$:

$$= \frac{\mathrm{P}\left(R(a_n) = 0 \mid \bigcap_{j=1}^{n} D_j = d_j\right)\mathrm{P}\left(\bigcap_{j=1}^{n+1} D_j = d_j\right)}{\mathrm{P}\left(R(a_n) = 0 \mid D_{n+1} = d_{n+1}\right)\mathrm{P}\left(D_{n+1} = d_{n+1}\right)}$$

$$= \left(\prod_{j=1}^{n}\psi_j\left(\prod_{i=j}^{n}\phi_i\right)^{d_j}\right)\frac{\Gamma(\alpha^{-1} + \sum_{j=1}^{n+1} d_j)}{\Gamma(\alpha^{-1})\prod_{j=1}^{n+1} d_j!}\left(\frac{1}{\mu_{n+1}}\right)^{\alpha^{-1}}\prod_{j=1}^{n+1}\left(\frac{\Delta\mu_j}{\mu_{n+1}}\right)^{d_j}$$

$$\times\left(\prod_{j=1}^{n}\psi_j\right)^{-1}\left(\frac{1 + \Delta\mu_{n+1}}{\mu_{n+1} - \sum_{j=1}^{n}\left(\prod_{i=j}^{n}\phi_i\right)\Delta\mu_j}\right)^{-(\alpha^{-1}+d_{n+1})}$$

$$\times\left(\frac{\Gamma(\alpha^{-1} + d_{n+1})}{\Gamma(\alpha^{-1})d_{n+1}!}\left(\frac{1}{1 + \Delta\mu_{n+1}}\right)^{\alpha^{-1}}\left(\frac{\Delta\mu_{n+1}}{1 + \Delta\mu_{n+1}}\right)^{d_{n+1}}\right)^{-1}$$

$$= \frac{\Gamma(\alpha^{-1} + \sum_{j=1}^{n+1} d_j)}{\Gamma(\alpha^{-1} + d_{n+1})\prod_{j=1}^{n} d_j!}\left(\frac{1}{\mu_{n+1}}\right)^{\alpha^{-1}}\prod_{j=1}^{n}\left(\frac{\left(\prod_{i=j}^{n}\phi_i\right)\Delta\mu_j}{\mu_{n+1}}\right)^{d_j}\left(\frac{\Delta\mu_{n+1}}{\mu_{n+1}}\right)^{d_{n+1}}$$

$$\times \left( \mu_{n+1} - \sum_{j=1}^{n} \left( \prod_{i=j}^{n} \phi_i \right) \Delta\mu_j \right)^{\alpha^{-1}} \left( \frac{\mu_{n+1} - \sum_{j=1}^{n} \left( \prod_{i=j}^{n} \phi_i \right) \Delta\mu_j}{\Delta\mu_{n+1}} \right)^{d_{n+1}}$$

$$= \frac{\Gamma(\alpha^{-1} + \sum_{j=1}^{n+1} d_j)}{\Gamma(\alpha^{-1} + d_{n+1}) \prod_{j=1}^{n} d_j!} \left( \frac{\mu_{n+1} - \sum_{j=1}^{n} \left( \prod_{i=j}^{n} \phi_i \right) \Delta\mu_j}{\mu_{n+1}} \right)^{\alpha^{-1} + d_{n+1}}$$

$$\prod_{j=1}^{n} \left( \frac{\left( \prod_{i=j}^{n} \phi_i \right) \Delta\mu_j}{\mu_{n+1}} \right)^{d_j}$$

The joint distribution of $\bigcap_{j=1}^{n} D_j \mid D_{n+1} = d_{n+1}, R(a_n) = 0$ is thus negative multinomial:

$$\bigcap_{j=1}^{n} D_j \mid D_{n+1} = d_{n+1}, R(a_n) = 0 \sim \mathcal{NM}\left( \alpha^{-1} + d_{n+1}, \left( \frac{\left( \prod_{i=j}^{n} \phi_i \right) \Delta\mu_j}{\mu_{n+1}} \right)_{j=1,\ldots,n} \right)$$

and, from proposition 2.3:

$$\sum_{j=1}^{n} D_j \mid D_{n+1} = d_{n+1}, R(a_n) = 0 \sim \mathcal{NB}\left( \alpha^{-1} + d_{n+1}, 1 - \frac{\sum_{j=1}^{n} \left( \prod_{i=j}^{n} \phi_i \right) \Delta\mu_j}{\mu_{n+1}} \right)$$

□

Passing to the limit ($n \to \infty$ while $\max_{j=1,\ldots,n} \left( a_j - a_{j-1} \right) \to 0+$), and setting in the above:

$$a_0 = 0, \quad a_n = a, \quad a_{n+1} = b,$$

yields the following proposition.

PROPOSITION 6.10.– The conditional distribution of $N(a-)$ given $N(b) - N(a)$ and $R(a-) = 0$ is negative binomial:

$$N(a-) \mid N(b) - N(a) = m, R(a-) = 0 \sim \mathcal{NB}\left( \alpha^{-1} + m, \frac{\mu(b) - \nu(a)}{\mu(b)} \right)$$

PROOF.–

The proposition is a straightforward consequence of proposition 6.9. □

7

# LEYP2s Likelihood and Inference

In order to be able to estimate LEYP2s parameters from actual failure and decommissioning data, we have to build the likelihood function of the parameters, as we did in section 5.1 in the case of the simple LEYP model. Within LEYP2s framework, we have to consider both failure $N(t)$ and decommissioning $R(t)$ processes conditioned on $R(a-)$, i.e. the survival of the considered object until the observation window $[a,b]$. The likelihood of joint processes $N(t)$ and $R(t)$ is defined as the product of the likelihood related to $N(t)$, the observation of which is censored by $R(t)$, by the likelihood related to $R(t)$, the intensity of which depends on $N(t)$.

DEFINITION 7.1.–

$$L(\theta) = \prod_{t\in[a,b]} \left\{ \mathrm{E}\left(\mathrm{d}N(t) \mid \mathcal{N}_{[a,t[}, R(a-)\right)^{\Delta N(t)} \left(1 - \mathrm{E}\left(\mathrm{d}N(t) \mid \mathcal{N}_{[a,t[}, R(a-)\right)\right)^{1-\Delta N(t)} \right.$$
$$\left. \times \mathrm{E}\left(\mathrm{d}R(t) \mid \mathcal{N}_{[a,t[}, R(a-)\right)^{\Delta R(t)} \left(1 - \mathrm{E}\left(\mathrm{d}R(t) \mid \mathcal{N}_{[a,t[}, R(a-)\right)\right)^{1-\Delta R(t)} \right\} \quad [7.1]$$

*with:*

$$\Delta N(t) = N(t) - N(t-), \quad \Delta R(t) = R(t) - R(t-)$$

Thus, we have to find closed analytical forms for both quantities:

$$\mathrm{E}\left(\mathrm{d}N(t) \mid N(t-) - N(a), R(a-) = 0\right)$$
$$\mathrm{E}\left(\mathrm{d}R(t) \mid N(t-) - N(a), R(a-) = 0\right)$$

We begin with the following proposition related to the conditional intensity of the failure process.

PROPOSITION 7.1.–

$$\mathrm{E}(\mathrm{d}N(t) \mid N(t-) - N(a) = m, R(a-) = 0) = (\alpha^{-1} + m)\mathrm{d}\ln(\mu(t) - \nu(a))$$

PROOF.–

Using the law of total probability:

$$\mathrm{E}(\mathrm{d}N(t) \mid N(t-) - N(a) = m, R(a-) = 0)$$

$$= \sum_{k=0}^{\infty} \mathrm{E}(\mathrm{d}N(t) \mid N(t-) - N(a) = m, R(a-) = 0, N(a-) = k)$$

$$\times \mathrm{P}(N(a-) = k \mid N(t-) - N(a) = m, R(a-) = 0)$$

$$= \sum_{k=0}^{\infty} (1 + \alpha(m+k)\lambda(t)\mathrm{d}t)\,\mathrm{P}(N(a-) = k \mid N(t-) - N(a) = m, R(a-) = 0)$$

$$= (1 + \alpha m + \alpha\mathrm{E}(N(a-) \mid N(t-) - N(a) = m, R(a-) = 0))\,\lambda(t)\mathrm{d}t$$

And using proposition 6.10 yields:

$$= (1 + \alpha m)\left(1 + \frac{\nu(a)}{\mu(t) - \nu(a)}\right)\lambda(t)\mathrm{d}t$$

$$= (1 + \alpha m)\frac{\mu(t)\lambda(t)\mathrm{d}t}{\mu(t) - \nu(a)}$$

$$= (\alpha^{-1} + m)\frac{\mathrm{d}\mu(t)}{\mu(t) - \nu(a)}$$

□

The following proposition relates to the conditional intensity of the decommissioning process.

PROPOSITION 7.2.–

$$\mathrm{E}(\mathrm{d}R(t) \mid N(t-) - N(a) = m, R(a-) = 0)$$

$$= \left(\psi(t) + \frac{\phi(\alpha m\mu(t) + \nu(a))}{\alpha(\mu(t) - \nu(a))}\right)\mathrm{d}t$$

PROOF.–

Using the law of total probability:

$$
\begin{aligned}
&\mathrm{E}\left(\mathrm{d}R(t) \mid N(t-) - N(a) = m, R(a-) = 0\right)\\
&= \sum_{k=0}^{\infty} \mathrm{E}\left(\mathrm{d}R(t) \mid N(t-) - N(a) = m, R(a-) = 0, N(a-) = k\right)\\
&\qquad \times \mathrm{P}\left(N(a-) = k \mid N(t-) - N(a) = m, R(a-) = 0\right)\\
&= \sum_{k=0}^{\infty} \left((\psi(t) + \phi(m+k))\,\mathrm{d}t\right) \mathrm{P}\left(N(a-) = k \mid N(t-) - N(a) = m, R(a-) = 0\right)\\
&= \left[\psi(t) + \phi m + \phi \mathrm{E}\left(N(a-) \mid N(t-) - N(a) = m, R(a-) = 0\right)\right] \mathrm{d}t
\end{aligned}
$$

And using proposition 6.10 yields:

$$
= \left(\psi(t) + \phi m + \phi(\alpha^{-1} + m)\frac{\nu(a)}{\mu(t) - \nu(a)}\right) \mathrm{d}t
$$

□

We are now able to establish an explicit analytical formula for LEYP2s likelihood, by applying propositions 7.1 and 7.2 in definition 7.1. To this end, we consider the likelihood related to a single technical object which is supposed to have been observed within age interval $[a, b]$, in which it underwent $m$ failures at ages $t_j, j = 1, \ldots, m$[1], and was at $b$ either still in service, i.e. $R(b) = 0$ (and obviously $\Delta R(b) = 0$ too), or decommissioned, i.e. $\Delta R(b) = 1$.

PROPOSITION 7.3.– The likelihood of the theoretical LEYP2s process with parameter $\boldsymbol{\theta}$ given $m$ observed failures within $[a, b]$ and possible decommissioning at $b$ is:

$$
\begin{aligned}
L(\boldsymbol{\theta}) = &\alpha^m \frac{\Gamma(\alpha^{-1} + m)}{\Gamma(\alpha^{-1})} \frac{(\mu(a) - \nu(a))^{\alpha^{-1}}}{(\mu(b) - \nu(a))^{\alpha^{-1}+m}} \prod_{j=1}^{m} \mu(t)_j \lambda(t_j)\\
&\times \left(\psi(b) + \frac{\phi(\alpha m \mu(b) + \nu(a))}{\alpha(\mu(b) - \nu(a))}\right)^{\Delta R(b)}
\end{aligned}
$$

1 By convention, $t_0 = a$ and $t_{m+1} = b$.

$$\times \exp\left( - \int_a^b \psi(t)\mathrm{d}t - \sum_{j=0}^{m} \int_{t_j}^{t_{j+1}} \frac{\phi(\alpha j \mu(t) + \nu(a))}{\alpha(\mu(t) - \nu(a))} \mathrm{d}t \right)$$

PROOF.–

For the first of the four terms within product integral in definition 7.1, related to $m$ failure times, we obtain:

$$\prod_{t \in [a,b]} \mathrm{E}\left(\mathrm{d}N(t) \mid \mathcal{N}_{[a,t[}, R(a-)\right)^{\Delta N(t)}$$

$$= \prod_{j=1}^{m} \mathrm{E}\left(\mathrm{d}N(t_j) \mid N(t_j-) - N(a), R(a-)\right)$$

$$= \prod_{j=1}^{m} (1 + \alpha(j-1)) \frac{\mu(t_j)\lambda(t_j)}{\mu(t_j) - \nu(a)}$$

$$= \alpha^m \frac{\Gamma(\alpha^{-1} + m)}{\Gamma(\alpha^{-1})} \prod_{j=1}^{m} \frac{\mu(t_j)\lambda(t_j)}{\mu(t_j) - \nu(a)}$$

For the second term, related to inter-failure times:

$$\prod_{t \in [a,b]} \left(1 - \mathrm{E}\left(\mathrm{d}N(t) \mid \mathcal{N}_{[a,t[}, R(a-)\right)\right)^{1 - \Delta N(t)}$$

$$= \prod_{j=0}^{m} \prod_{t \in [t_j, t_{j+1}[} \left(1 - \mathrm{E}\left(\mathrm{d}N(t) \mid N(t-) - N(a), R(a-)\right)\right)$$

$$= \prod_{j=0}^{m} \exp\left( - \int_{t_j}^{t_{j+1}} (\alpha^{-1} + j) \mathrm{d} \ln(\mu(t) - \nu(a)) \right)$$

$$= \prod_{j=0}^{m} \left( \frac{\mu(t_j) - \nu(a)}{\mu(t_{j+1}) - \nu(a)} \right)^{\alpha^{-1} + j}$$

$$= \frac{(\mu(a) - \nu(a))^{\alpha^{-1}}}{(\mu(b) - \nu(a))^{\alpha^{-1} + m}} \prod_{j=1}^{m} \left(\mu(t_j) - \nu(a)\right)$$

For the third term, related to the case where the object is decommissioned at $b$:

$$\prod_{t\in[a,b]} \mathrm{E}\,(\mathrm{d}R(t) \mid \mathcal{N}_{[a,t[}, R(a-))^{\Delta R(t)}$$

$$= \mathrm{E}\,(\mathrm{d}R(b) \mid N(b-) - N(a), R(a-))^{\Delta R(b)}$$

$$= \left(\psi(b) + \frac{\phi(\alpha m\mu(b) + \nu(a))}{\alpha(\mu(b) - \nu(a))}\right)^{\Delta R(b)}$$

And for the fourth term, related to the case where the object is still in service at $b$:

$$\prod_{t\in[a,b]} (1 - \mathrm{E}\,(\mathrm{d}R(t) \mid \mathcal{N}_{[a,t[}, R(a-)))^{1-\Delta R(t)}$$

$$= \prod_{j=0}^{m} \prod_{t\in[t_j,t_{j+1}[} (1 - \mathrm{E}\,(\mathrm{d}R(t) \mid N(t-) - N(a), R(a-)))$$

$$= \prod_{j=0}^{m} \exp\left(-\int_{t_j}^{t_{j+1}} \left(\psi(t) + \frac{\phi(\alpha j\mu(t) + \nu(a))}{\alpha(\mu(t) - \nu(a))}\right) \mathrm{d}t\right)$$

$$= \exp\left(-\int_a^b \psi(t)\mathrm{d}t - \sum_{j=0}^{m} \int_{t_j}^{t_{j+1}} \frac{\phi(\alpha j\mu(t) + \nu(a))}{\alpha(\mu(t) - \nu(a))}\mathrm{d}t\right)$$

□

For practical applications, we will consider, as in section 5.1, the log-likelihood related to a collection of $n$ objects, assumed to undergo independent failure and decommissioning processes:

$$\begin{aligned}\ln L(\boldsymbol{\theta}) = \sum_{i=1}^{n} & m_i \ln\alpha + \ln\Gamma(\alpha^{-1} + m_i) - \ln\Gamma(\alpha^{-1}) \\ & + \alpha^{-1}\ln(\mu(a_i) - \nu(a_i)) - (\alpha^{-1} + m_i)\ln(\mu(b_i) - \nu(a_i)) \\ & + I_{[m_i>0]} \sum_{j=1}^{m_i} \left(\ln\lambda(t_{ij}) + \alpha\Lambda(t_{ij})\right) \\ & + \Delta R(b_i) \ln\left(\psi(b_i) + \frac{\phi(\alpha m_i\mu(b_i) + \nu(a_i))}{\alpha(\mu(b_i) - \nu(a_i))}\right)\end{aligned}$$

$$-\int_{a_i}^{b_i}\psi(t)\mathrm{d}t-\sum_{j=0}^{m_i}\int_{t_{ij}}^{t_{ij+1}}\frac{\phi(\alpha j\mu(t)+\nu(a_i))}{\alpha(\mu(t)-\nu(a_i))}\mathrm{d}t \qquad [7.2]$$

## 7.1. Validation of the estimation procedure for LEYP2s

The log-likelihood computation formula [7.2] looks to be somewhat complicated, and this raises the question of whether an optimal value of the parameter $\boldsymbol{\theta}$ can be easily found or not, by using the Nelder–Mead optimization procedure (see section 5.2.5). As for many statistical estimation problems, this question does not seem to be easily answered from a theoretical point of view. The practical implementation of LEYP2s seems nevertheless to deliver interesting results, which will be illustrated in Chapter 8. In order to gain some confidence in our estimation procedure, we will carry out the following steps on a theoretical example:

– generate a pseudo-random failure and decommissioning dataset, according to a given LEYP2s parameter;

– check parameter estimates with respect to theoretical values;

– check convexity of log-likelihood function in a reasonably broad neighborhood around the estimates.

### 7.1.1. *Constrained and selective decommissioning survival functions*

Until now, the function $\psi(t)$ that drives constrained decommissioning has been left unspecified in LEYP2s in definition 6.1. We will see later in Chapter 8 that the Weibull hazard function turns out to be a good choice:

$$\psi(t)=\psi_0\psi_1 t^{\psi_1-1} \qquad [7.3]$$

with $\psi_0 \geqslant 0$ and $\psi_1 \geqslant 0$.

The constrained decommissioning survival function is then:

$$S_C(t)=\mathrm{P}\left(R_C(t)=0\right)=\exp\left(-\psi_0 t^{\psi_1}\right) \qquad [7.4]$$

The selective decommissioning survival function conditional to failure times can be directly deduced from definition 6.1, with $T_j$ standing as usual for the $j$th failure time, and by convention $t_{N(t-)+1} = t$:

$$S_S(t \mid N(s)_{s<t}) = \mathrm{P}\left(R_S(t) = 0 \mid T_j = t_j, j = 1, \ldots, N(t-)\right)$$
$$= \exp\left(-\sum_{j=1}^{N(t-)} j\phi\left(t_{j+1} - t_j\right)\right) \qquad [7.5]$$

### 7.1.2. *Random failure and decommissioning data generation*

The random simulation of failure and decommissioning events consists first of setting theoretical parameter values. The values displayed in Table 7.1 have been chosen because they are realistic with respect to what can be observed for water network segments (see Chapter 8). Regression parameters $\beta_1, \beta_2$ and $\beta_3$, respectively, mimic the effect of covariates $Z_1$ = log-transform of segment length (m), $Z_2$ = pipe diameter (mm) and $Z_3$ an indicator variable for segment location under roadways ($Z_3 = 1$) versus under sidewalks ($Z_3 = 0$).

| Label | Value |
|---|---|
| $\psi_0$ | 0.8 |
| $\psi_1$ | 2.0 |
| $\phi$ | 2.0 |
| $\alpha$ | 3.0 |
| $\delta$ | 1.3 |
| $\beta_0$ | -2.2 |
| $\beta_1$ | 0.5 |
| $\beta_2$ | -0.0024 |
| $\beta_3$ | 0.2 |

**Table 7.1.** *LEYP2s theoretical parameter values*

The installation of 10,000 network segments is simulated as being uniformly distributed over the years 1870–1970, whereas the observation window spans the years 1985–2015 (i.e. $D_1$ = 1985 and $D_2$ = 2015). Segment lengths (log-transformed into covariate $Z_1$) follow a truncated exponential distribution between 20 m and 500 m with an expectation of 30 m. Pipe diameters (covariate $Z_2$) follow a truncated exponential distribution between 100 mm and 1,000 mm, discretized by a 100 mm step, with an expectation of 150 mm. Segment location under roadways (covariate $Z_3$) is randomly set to 1 with a probability of 0.5.

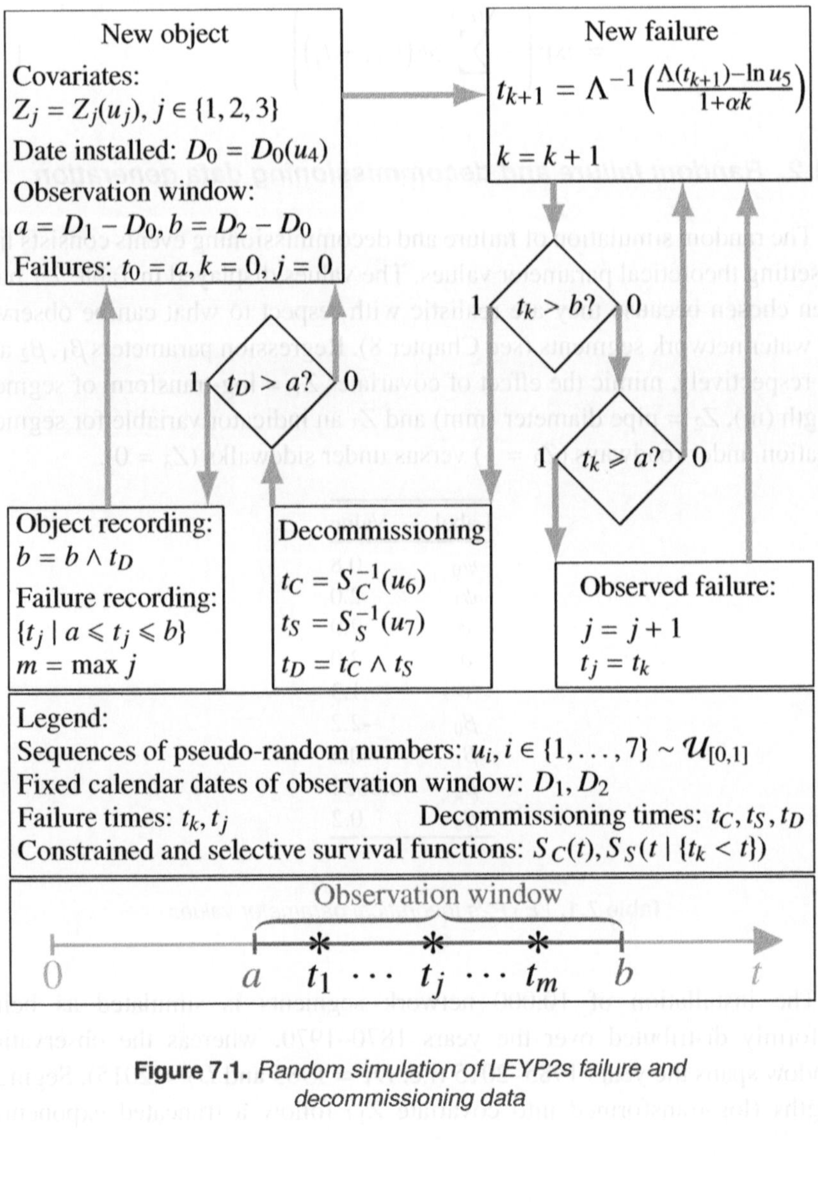


**Figure 7.1.** *Random simulation of LEYP2s failure and decommissioning data*

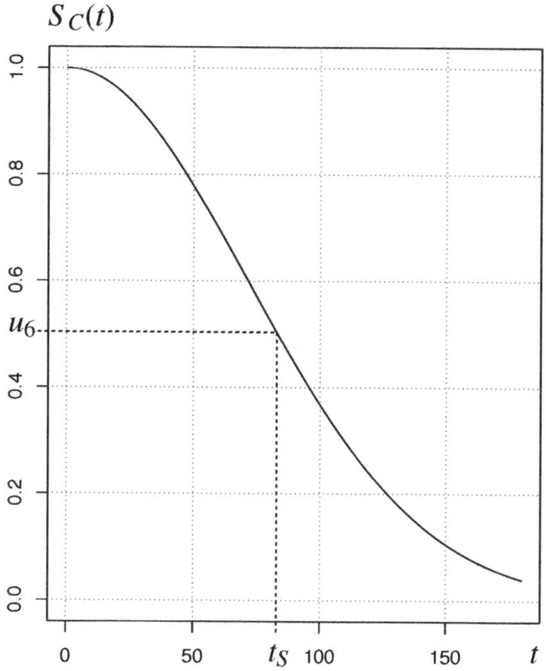


**Figure 7.2.** *Constrained decommissioning survival curve*

The simulation algorithm is shown in Figure 7.1. Four main steps can be distinguished:

– the first step intitializes a new segment and randomly valuates its installation year $D_0$ (and then observation ages $a$ and $b$), as well as its length, diameter and location;

– the second step relates to simple LEYP failure simulations, which are stopped as soon as the $k$th failure time exceeds $b$;

– the third step simulates the age $t_D$ at decommissioning, which is the minimum of both random ages at decommissioning $t_C$ for either constraint/opportunity reason, or $t_S$ motivated by repeated failures; the constrained and selective decommissioning survival functions $S_C(t)$ and $S_S(t)$ are, illustrated by Figures 7.2 and 7.3, respectively;

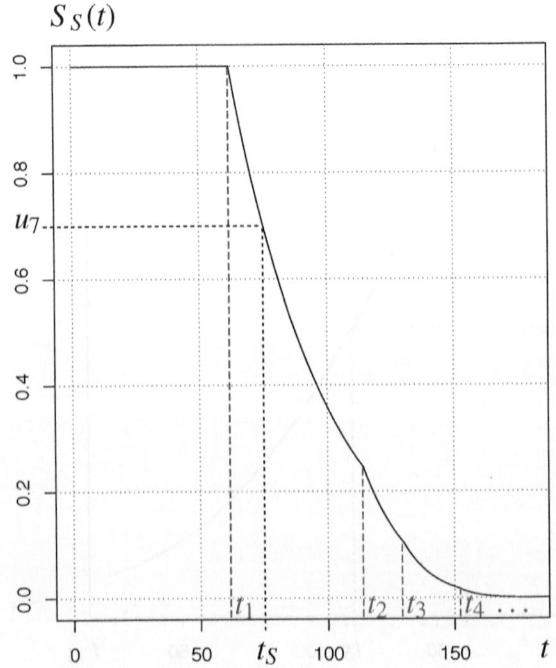


**Figure 7.3.** *Selective decommissioning survival curve*

– the fourth step discards segments decommissioned before $a$ (segment left-truncation) and, concerning segments kept, discards failures undergone before $a$ (failure left-truncation) and failures undergone beyond $b$ (failure right-censoring).

Around 4,000 segments are decommissioned, due to random left-truncation. The average simulated length-weighted failure rate within the observation window is 0.185 $km^{-1}.yr^{-1}$. The distribution of the simulated number of failures per segment $m$ is displayed in Table 7.2, which illustrates the high failure concentration on a few segments, typical of LEYP, and more generally of negative binomial distribution. The average simulated length-weighted decommissioning rate is 0.016 yr$^{-1}$, 2,283 segments being decommissioned before 2015, i.e. right-censored.

| $m$ | 0 | 1 | 2 | 3 | 4 | 5 | 6 | 7 |
|---|---|---|---|---|---|---|---|---|
| Count | 4 834 | 762 | 216 | 85 | 41 | 24 | 4 | 3 |

**Table 7.2.** *Distribution of the simulated number of failures per segment*

### 7.1.3. *Checking parameter estimate accuracy*

After convergence of Nelder–Mead optimization algorithm, the estimation results displayed in Table 7.3 are obtained. All estimates are significantly different from the reference values, which, respectively, characterize:

– no constrained decommissioning for $\psi_0 = 0$;

– constrained decommissioning independent of age for $\psi_1 = 1$;

– no selective decommissioning for $\phi = 0$;

– NHPP-like behaviour for $\alpha = 0$;

– absence of ageing for $\delta = 1$;

– absence of covariate effects for $\beta_1 = 0$, $\beta_2 = 0$ and $\beta_3 = 0$ (the test is meaningless concerning $\beta_0$ which is a simple scaling factor).

| Label | Estimate | Std. Dev. | Ref. | Chi2 (DF) | P-Value |
|---|---|---|---|---|---|
| $\psi_0$ | 7.7199e-01 | 1.3674e-02 | 0 | 3.1789e+03 (1) | 0.0000 |
| $\psi_1$ | 2.0088e+00 | 6.2371e-02 | 1 | 2.6158e+02 (1) | 0.0000 |
| $\phi$ | 1.9680e+00 | 5.3852e-02 | 0 | 1.3342e+03 (1) | 0.0000 |
| $\alpha$ | 3.1160e+00 | 7.4961e-02 | 0 | 1.7268e+03 (1) | 0.0000 |
| $\delta$ | 1.2089e+00 | 3.7422e-02 | 1 | 3.1175e+01 (1) | 0.0000 |
| $\beta_0$ | -2.2007e+00 | 1.5447e-02 | 0 | 2.0298e+04 (1) | 0.0000 |
| $\beta_1$ | 4.8094e-01 | 6.7503e-03 | 0 | 5.0763e+03 (1) | 0.0000 |
| $\beta_2$ | -2.3200e-03 | 8.8314e-05 | 0 | 6.9009e+02 (1) | 0.0000 |
| $\beta_3$ | 1.9667e-01 | 2.8419e-02 | 0 | 4.7894e+01 (1) | 0.0000 |

**Table 7.3.** *Parameter estimates and significance tests for LEYP2s model – synthetic data*

Note that the estimated standard-deviations of parameter estimates are quite small compared to estimate magnitudes; this is attributable both to the accuracy of the estimation procedure and to the high sample size used. Table 7.4 shows that all estimates are close to their theoretical counterparts, which are most often included in the confidence interval of estimates or very close to it.

| Label | Theoretical | Estimate | Lower bound | Upper bound |
|---|---|---|---|---|
| $\psi_0$ | 0.8 | 0.772 | 0.745 | 0.799 |
| $\psi_1$ | 2.0 | 2.009 | 1.887 | 2.131 |
| $\phi$ | 2.0 | 1.968 | 1.862 | 2.074 |
| $\alpha$ | 3.0 | 3.116 | 2.969 | 3.263 |
| $\delta$ | 1.3 | 1.209 | 1.136 | 1.282 |
| $\beta_0$ | -2.2 | -2.201 | -2.231 | -2.170 |
| $\beta_1$ | 0.5 | 0.481 | 0.468 | 0.494 |
| $\beta_2$ | -0.0024 | -0.0023 | -0.0025 | -0.0021 |
| $\beta_3$ | 0.2 | 0.197 | 0.141 | 0.252 |

**Table 7.4.** *LEYP2s parameter estimates, with 95% confidence interval, compared to theoretical values – synthetic data*

REMARK 7.1.– This kind of numerical simulations has been repeatedly carried out by the author, and with various theoretical parameters, which always happen to be reasonably well estimated by log-likelihood maximization. This leads better confidence in the computational model estimation procedure. △

### 7.1.4. *Checking log-likelihood convexity*

The random simulation has also been used to graphically assess the shape of the log-likelihood function. Figure 7.4 displays one graph for each parameter, obtained by making its value vary in a rather broad neighborhood around its estimate, and setting all other parameter values at their estimated value. The graphs allow us to check the apparent convexity of the log-likelihood function, which explains the accuracy and rather quick convergence of the estimation procedure. This also validates the relevance of the box-constrained adaptation implemented in the Nelder–Mead algorithm to ensure the constraints $\psi_0 \geqslant 0$, $\psi_1 \geqslant 1$, $\phi \geqslant 0$, $\alpha > 0$ and $\delta \geqslant 1$ are matched.

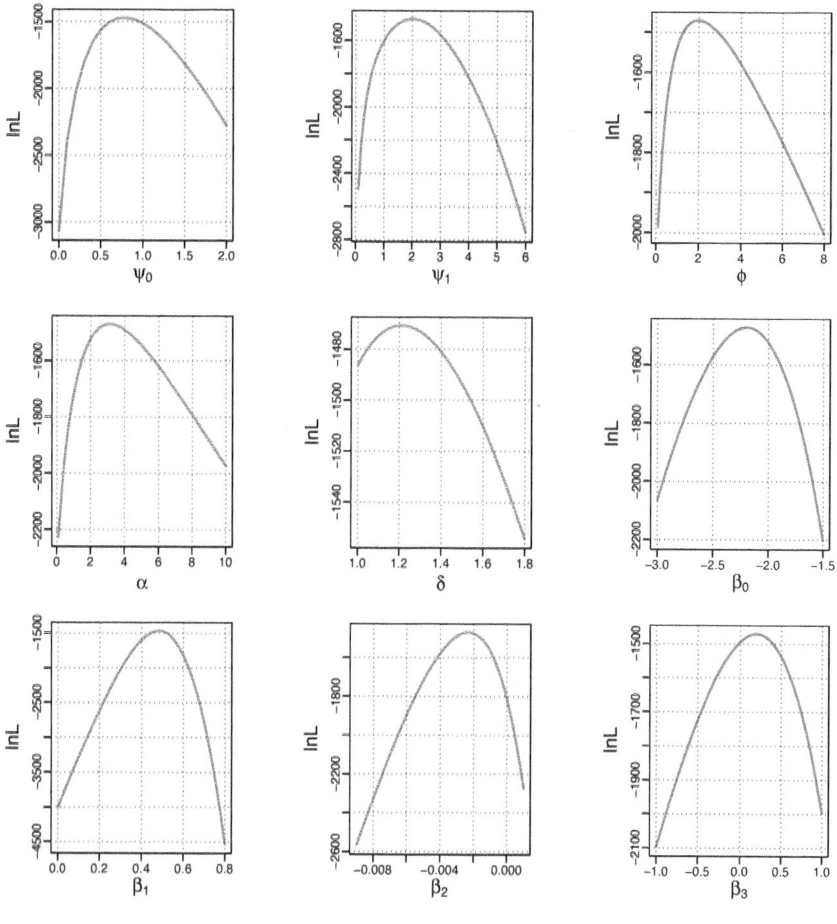

**Figure 7.4.** *LEYP2s log-likelihood shape around theoretical parameter values*

# 8

# Case Study Application of the LEYP2s Model

## 8.1. Lausanne water utility

Lausanne is a medium-sized European city, located on the edge of Lake Geneva in Switzerland. The water utility *eauservice* is in charge of the drinking water network, a public company controlled by the Lausanne municipality.

Due to the technical level of its staff, *eauservice* keeps the Lausanne water network in good operational condition, and supplies a high-quality service, in matters of water quality, service continuity and pressure, to the 360,000 inhabitants of its 17 municipalities. *Eauservice* has for a long time emphasized its will to keep its network management method at the cutting edge of technological and scientific developments, which particularly led, in the beginning of 2000, to its collaboration within the CARE-W European project, which aimed at aiding water network rehabilitation decisions (see [HER 03] and [SAE 05]).

Collaborating with *eauservice* we were able to gain access to high-quality technical data, which is illustrated in this chapter.

## 8.2. Lausanne water supply network

The Lausanne water network is 721 km long, mainly composed of 93 km in gray cast iron (GCI), and 515 km in ductile iron (DI); these figures are from 2013, and only concern the supply network (transport pipelines are excluded).

As illustrated by Figures 8.1 and 8.2, respectively, GCI pipes were installed between 1870 and 1969, and DI pipes since 1965.

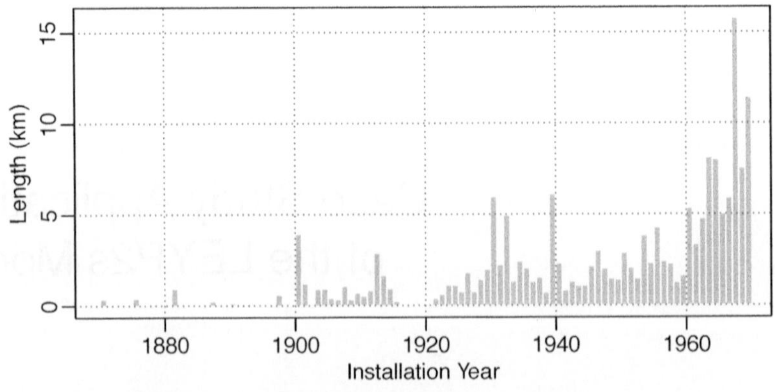


**Figure 8.1.** *Installation year distribution of Lausanne gray cast iron pipes*

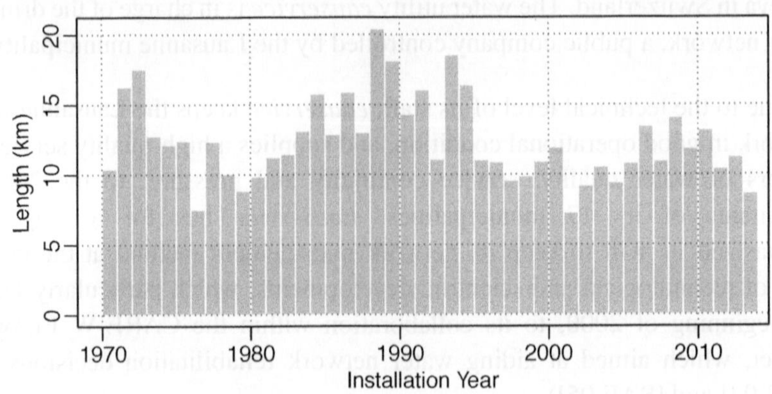


**Figure 8.2.** *Installation year distribution of Lausanne ductile iron pipes*

The network is precisely described by geographical information system databases, which are carefully updated on a daily basis. The network is subdivided into *segments*, which are defined as pipelines homogeneous with respect to their installation year, material and diameter, and most often bounded by valves. Note that segment length can vary, and typically ranges from 5 to 500 m.

A water network is a burried infrastructure, and this feature is particularly important regarding the reliability study of its pipes, as these components are not directly observable. Failure occurrences therefore, play a pivotal role.

## 8.3. Lausanne network segment failure and decommissioning data

All failure (pipe leaks or breaks that led to repairs) and decommissioning (pipe structural rehabilitations, replacements or relocations) events are reported on a daily basis, and resulting data can be made available in electronic format for statistical studies. The data processed for the purpose of this book span from January 1st 2001 to December 31st 2013, and are exhaustive with respect to the network segments. The present study focuses on GCI and DI pipes.

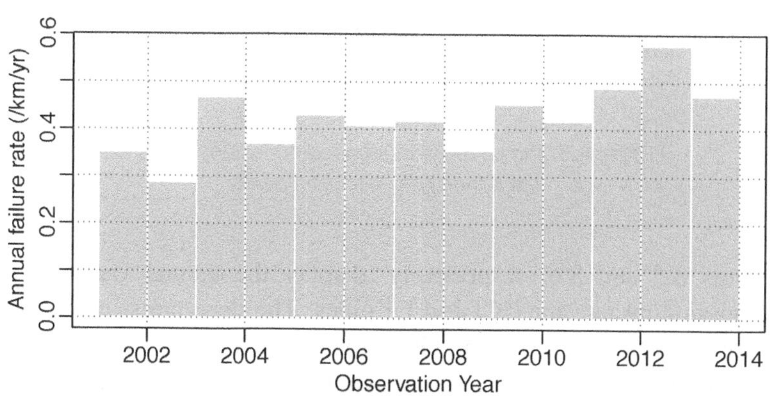


**Figure 8.3.** *Observed annual failure rate of Lausanne gray cast iron pipes*

Figures 8.3 and 8.4, respectively, display the annual variation of the failure rate for GCI and DI pipes. The comparison of both graphs suggests that the average failure rate of DI pipes (0.102 /km/year) is four times less than that of GCI pipes (0.414 /km/year). The LEYP2s modeling of failure intensity in section 8.4 will provide some elements of explanation, especially concerning the respective effects of the nature of material, aging and past failures.

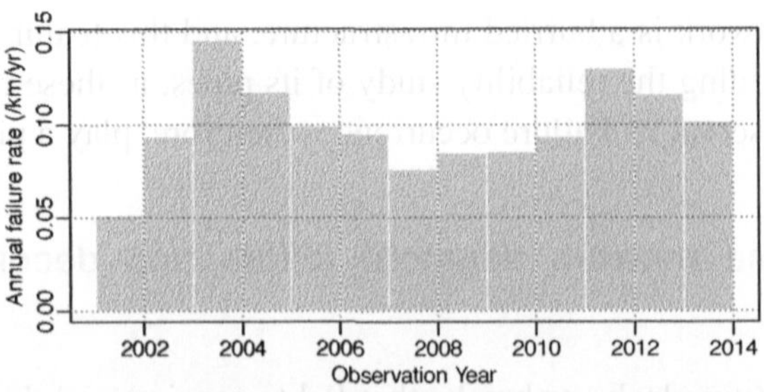


**Figure 8.4.** *Observed annual failure rate of Lausanne ductile iron pipes*

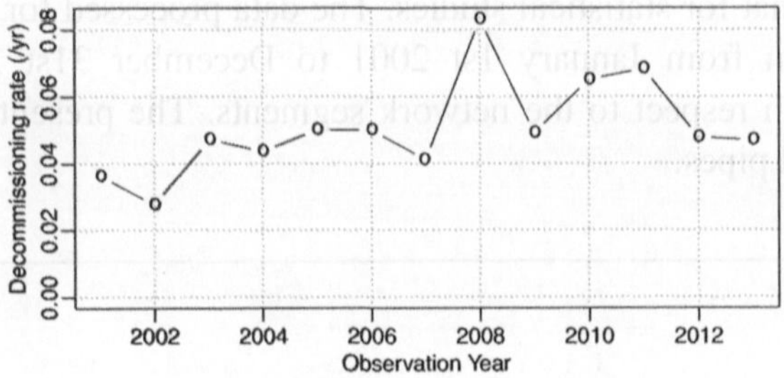


**Figure 8.5.** *Observed annual decommissioning rate of Lausanne gray cast iron pipes*

Figures 8.5 and 8.6, respectively, display the annual variation of the decommissioning rate for GCI and DI pipes. The decommissioning pressure is apparently 10 times higher on GCI pipes (0.052/year versus 0.0055/year), which reflects the *eauservice* proactive policy of GCI pipe elimination. Roadwork opportunities are indeed systematically used to replace GCI segments by DI segments. This is a key point to keep in mind when considering the GCI survival curves in section 8.7.

## 8.4. Model parameter estimates

LEYP2s model parameters have been estimated separately for GCI and DI network segments, the failure and decommissioning events of which were

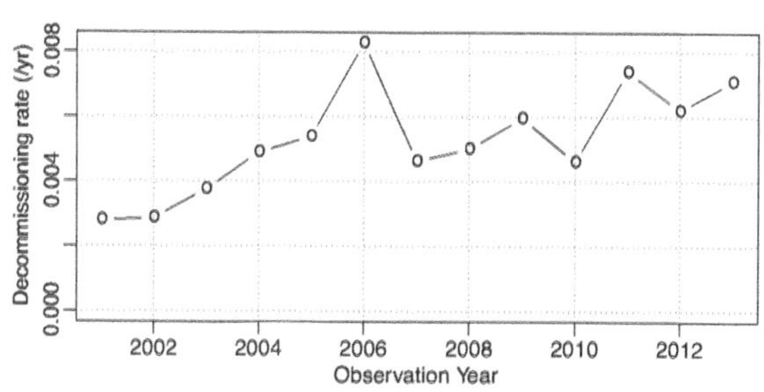


**Figure 8.6.** *Observed annual decommissioning rate of Lausanne ductile iron pipes*

| Material | Segments | Failures | Decommissionings |
|---|---|---|---|
| GCI | 2,248 | 685 | 1,269 |
| DI | 7,071 | 598 | 719 |

**Table 8.1.** *Numbers of observed segments, failure and decommissioning events per material – Lausanne water utility data 2001–2013*

exhaustively recorded from 2001 to 2013. Table 8.3 displays the numbers of observed segments, failure and decommissioning events for both materials.

The description of network segments in the Lausanne water utility database contains exhaustive information about their length (m), pipe diameter (mm), service pressure (bar), presence of internal lining (0/1), presence of supply connections (0/1) and presence of cathodic protection (0/1). This information was introduced as covariates into the Cox factor of the LEYP2s intensity function, as presented in section 5.2.3. LEYP2s parameter estimates are presented in Tables 8.3 and 8.4, respectively, both for materials GCI and DI.

Cathodic protection was never found to be significant, whereas the presence of supply connections was significant for sole GCI model, and internal lining for sole DI model.

| Label | Estimate | Std. Dev. | Ref. | Chi2 (DF) | P-Value |
|---|---|---|---|---|---|
| $\psi_0$ | 3.9104e+00 | 7.6773e-02 | 0 | 2.5930e+03 (1) | 0.0000 |
| $\psi_1$ | 2.1912e+00 | 9.5288e-02 | 1 | 1.5628e+02 (1) | 0.0000 |
| $\phi$ | 2.7171e+00 | 1.1084e-01 | 0 | 6.0050e+02 (1) | 0.0000 |
| $\alpha$ | 2.0662e+00 | 8.4924e-02 | 0 | 5.9138e+02 (1) | 0.0000 |
| $\delta$ | 1.1367e+00 | 7.5763e-02 | 1 | 3.2537e+00 (1) | 0.0713 |
| Intercept | -1.3968e+00 | 2.5825e-02 | 0 | 2.9255e+03 (1) | 0.0000 |
| ln(Length) | 3.2762e-01 | 8.9087e-03 | 0 | 1.3524e+03 (1) | 0.0000 |
| Diameter | -2.3732e-03 | 1.8574e-04 | 0 | 1.6325e+02 (1) | 0.0000 |
| Pressure | 6.8424e-02 | 3.8392e-03 | 0 | 3.1764e+02 (1) | 0.0000 |
| Supply | 1.3727e-01 | 3.1604e-02 | 0 | 1.8866e+01 (1) | 0.0000 |

**Table 8.2.** *Parameter estimates (calibration years 2001–2013) and significance tests for LEYP2s model – Lausanne gray cast iron data*

| Label | Estimate | Std. Dev. | Ref. | Chi2 (DF) | P-Value |
|---|---|---|---|---|---|
| $\psi_0$ | 2.0977e+00 | 3.4270e-02 | 0 | 3.7431e+03 (1) | 0.0000 |
| $\psi_1$ | 1.9658e+00 | 2.9957e-02 | 1 | 1.0395e+03 (1) | 0.0000 |
| $\phi$ | 2.5662e+00 | 1.9831e-01 | 0 | 1.6732e+02 (1) | 0.0000 |
| $\alpha$ | 2.3481e+00 | 1.2497e-01 | 0 | 3.5276e+02 (1) | 0.0000 |
| $\delta$ | 3.0480e+00 | 2.3857e-02 | 1 | 7.3697e+03 (1) | 0.0000 |
| Intercept | 3.1926e-01 | 3.9733e-02 | 0 | 6.4560e+01 (1) | 0.0000 |
| ln(Length) | 3.8558e-01 | 1.1129e-02 | 0 | 1.2003e+03 (1) | 0.0000 |
| Diameter | -5.1757e-03 | 1.7491e-04 | 0 | 8.7562e+02 (1) | 0.0000 |
| Pressure | 3.8318e-02 | 4.1365e-03 | 0 | 8.5809e+01 (1) | 0.0000 |
| Lining | 1.1606e+00 | 1.1829e-01 | 0 | 9.6265e+01 (1) | 0.0000 |

**Table 8.3.** *Parameter estimates (calibration years 2001–2013) and significance tests for LEYP2s model – Lausanne ductile iron data*

Segment length was found to significantly influence the failure intensity, but raised at power between 0.3 and 0.4; this result is compliant with what is usually observed in such studies (see [LEG 00] or [LEG 14]). We should *a priori* intuitively expect a parameter estimate close to 1, but this does not happen because segment length also carries information about pipe environment, as segments installed in a dense city center are generally shorter, and more subject to environmental stresses (e.g. vibrations) than these installed in peripheral areas.

Larger pipe diameter tends to significantly reduce the failure intensity for two reasons:

– large diameter pipes are designed with a thicker pipe wall than smaller diameter pipes;

– the mechanical behavior of large diameter pipes tends to be that of a ring, instead of that of a beam for smaller diameters, hence more prone to circular breaks when subjected to differential settlements in their embedding soil.

High-service pressure is logically found to significantly increase failure intensity for both GCI and DI pipes.

Pipe aging has a marked impact for both materials, as $\delta$ parameter is always found to be significantly greater than 1. This effect appears to be, however, much stronger for DI segments, which is undoubtedly due to an observation bias: all observable GCI pipes are obviously much older, and likely to have already been selected on their robustness, whereas this selection process is still in progress for observable DI pipes. This consideration illustrates a limit of LEYP2s model to completely correct the selective survival bias, when based on a short and recent observation window, which is always the case when studying infrastructures that were partially installed a very long time ago.

For both materials, the Yule parameter $\alpha$ is found to be close to 2 or slightly higher, which characterizes the significant tendency of pipe failures to concentrate on a minority of segments. This makes the use of the LEYP2s model very efficient in helping to implement selective decommissioning of pipes. This usage may nevertheless have the paradoxical effect on the long term to increase the weight of the selective survival bias in the failure data, and hence will make model calibration more difficult. This effect is apparent in the reduced predictive power of the GCI model compared to that of DI, illustrated by Figures 8.9 and 8.10, presented in section 8.6.

Parameter $\phi$ obtains close estimates for both GCI ($\phi = 2.7$) and DI ($\phi = 2.6$) models, which suggests similar selective decommissioning policies for both materials. In the matter of constrained decommissioning, parameter estimates $\psi_1$ are close to 2, but $\psi_0$ estimate for GCI ($\psi_0 = 3.9$) is almost twice that of DI ($\psi_0 = 2.1$), which confirms the voluntarist policy of GCI elimination already mentioned above in section 8.3.

## 8.5. Model goodness of fit assessment

The comparison, in Table 8.5, for both GCI and DI models, of the total number of failures observed within the calibration window (2001–2013) to the expected ones, accompanied with their 95% confidence intervals (see section 5.5.2), does not reveal any significant bias.

| Material | Observed failures | Expected failures | Expected CI95% |
|---|---|---|---|
| GCI | 685 | 731.4 | [652.7, 810.1] |
| DI | 598 | 596.9 | [528.6, 665.2] |

**Table 8.4.** *Number of failures between 2001 and 2013, observed versus expected, with 95% confidence interval*

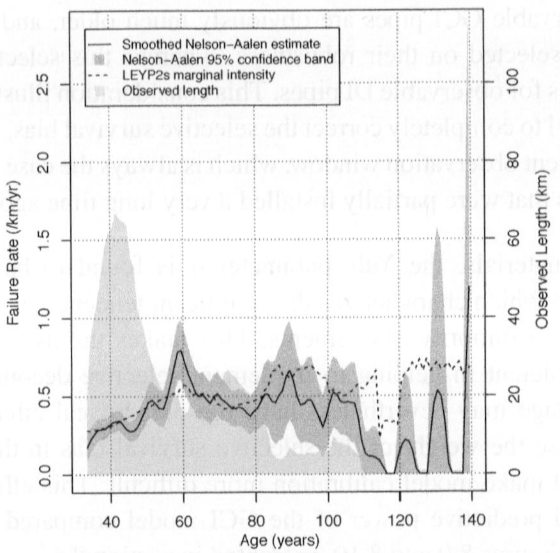


**Figure 8.7.** *Empirical failure rate versus LEYP2s intensity – Lausanne gray cast iron data*

Empirical versus expected failure rates are compared using the method presented in section 5.4, and the results are displayed in Figures 8.7 and 8.8. The goodness of fit appears to be satisfactory, except for GCI segments whose age exceeds 100 years, and for which the observable sample size is much reduced.

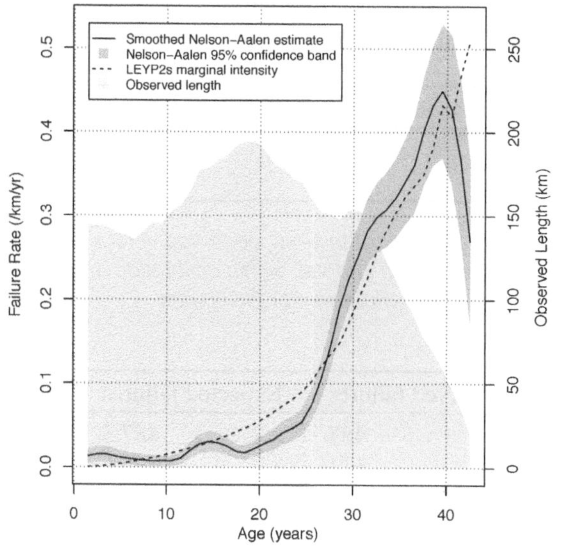


**Figure 8.8.** *Empirical failure rate versus LEYP2s intensity – Lausanne ductile iron data*

## 8.6. Model validation

GCI and DI LEYP2s models are validated using the method presented in section 5.5. The models are calibrated using the reduced observation window (2001–2011), and model predictions are compared to observations within the validation window (2012–2013). The resulting Lorenz curves are displayed in Figures 8.9 and 8.10. As explained in section 8.4, the predictive efficiency appears to be somewhat weaker in the case of GCI model compared to DI model, for which the area under Lorenz curve is quite high (0.88).

The same bias assessment as in section 8.5 has been carried out, with the results displayed in Tables 8.5 and 8.6. Predictions appear to be with slightly underestimated in the GCI case, and slightly overestimated for the DI model. These biases may be due to the reduced number of available validation years, or due to climatic effects that generate interannual fluctuations in the failure intensity, as suggested by Figures 8.3 and 8.4. A possible solution to improve the model robustness lies in time-dependent covariates introduction in the LEYP2s framework, which will be discussed further in Chapter 9.

| Window | Observed failures | Expected failures | Expected CI95% |
|---|---|---|---|
| 2001–2011 | 589 | 619.1 | [548.6, 689.7] |
| 2012–2013 | 96 | 75.5 | [57.5, 93.5] |

**Table 8.5.** *Number of failures, observed versus expected, with 95% confidence interval, within calibration and validation windows, for GCI pipes*

| Window | Observed failures | Expected failures | Expected CI95% |
|---|---|---|---|
| 2001–2011 | 487 | 487.3 | [426.6, 547.9] |
| 2012–2013 | 111 | 136.6 | [112.5, 160.7] |

**Table 8.6.** *Number of failures, observed versus expected, with 95% confidence interval, within calibration and validation windows, for DI pipes*

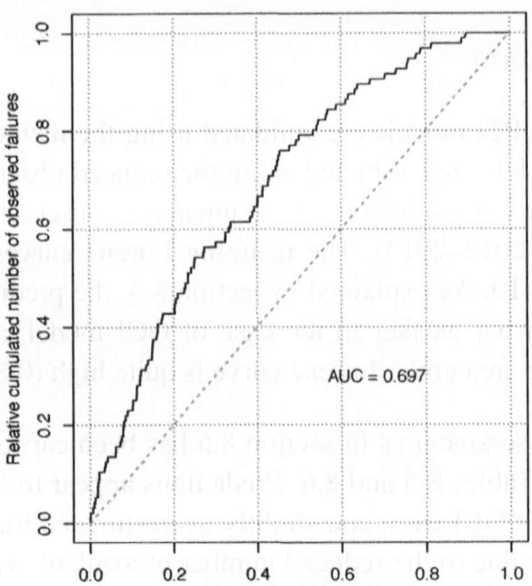


**Figure 8.9.** *Lorenz curve related to LEYP2s model calibration years 2001–2011 versus validation years 2012–2013 – Lausanne gray cast iron data*

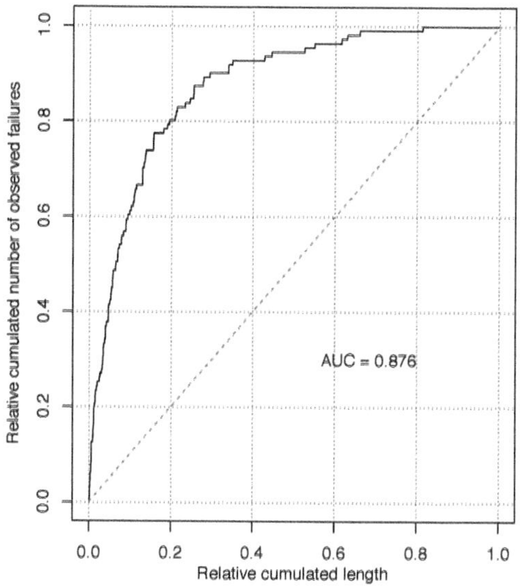


**Figure 8.10.** *Lorenz curve related to LEYP2s model calibration years 2001–2011 versus validation years 2012–2013 – Lausanne ductile iron data*

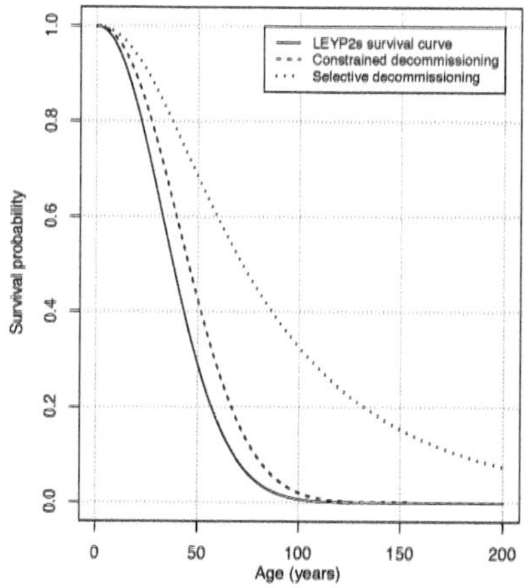


**Figure 8.11.** *Decomposition of LEYP2s survival into constrained and selective parts – Lausanne gray cast iron data*

## 8.7. Service lifetime

As explained in sections 6.2 and 7.1.2, an important feature of LEYP2s lies in its ability to model the survival function as the product of a constrained survival function, given by [7.4], and a selective survival function, given by [7.5]. This decomposition is illustrated by Figures 8.11 and 8.12. GCI and DI are not observed in the same range of ages (beyond 50 years for GCI, below 50 years for DI), but constrained decommissioning seems to prevail in both cases; extrapolating Figure 8.12 suggests that selective and constrained decommissioning could become more balanced, but this concerns DI pipe ages that are not yet observed.

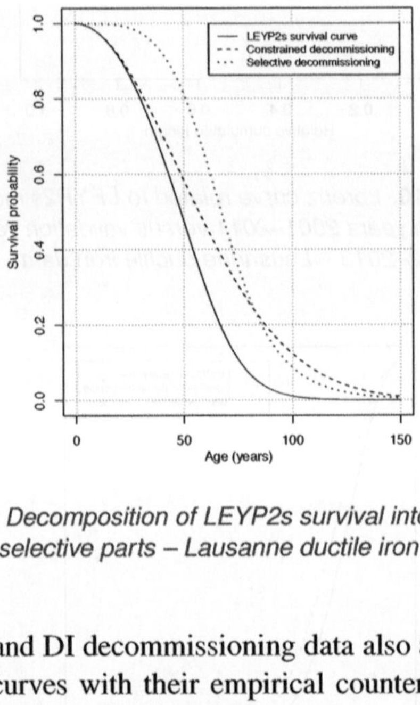


**Figure 8.12.** *Decomposition of LEYP2s survival into constrained and selective parts – Lausanne ductile iron data*

Available GCI and DI decommissioning data also allow us to compare the LEYP2s survival curves with their empirical counterparts, estimated by the Kaplan–Meier method (see [AND 93]), which is adapted to take the left-truncation into account. Using similar notations as in [7.2], the Kaplan–Meier survival curve is calculated as:

$$\hat{S}_{KM}(t) = \prod_{s \leqslant t} \left(1 - \frac{d_s}{Y_s}\right)$$

with:

$$d_s = \#\{i : b_i = s, \Delta R(b_i) = 1\}$$

$$Y_s = \#\{i : a_i \leqslant s \leqslant b_i\}$$

The variance of $\hat{S}_{KM}(t)$ can be estimated by the Greenwood's formula:

$$\widehat{\mathrm{Var}}\hat{S}_{KM}(t) = \hat{S}_{KM}(t)^2 \sum_{s \leqslant t} \frac{d_s}{Y_s(Y_s - d_s)}$$

Provided that the sample size is large enough for each observed decommissioning time, a 95% confidence interval is given by:

$$\hat{S}_{KM}(t)^{\exp\left(\frac{\pm 1.96 \widehat{\mathrm{Var}}\hat{S}_{KM}(t)^{1/2}}{\hat{S}_{KM}(t) \ln \hat{S}_{KM}(t)}\right)}$$

The Kaplan–Meier survival curves, accompanied by their 95% confidence bands, are compared in Figures 8.13 and 8.14 with their expected LEYP2s counterparts. The fit is reasonably good, which allows us to place some trust in the relevance of the decommissioning model.

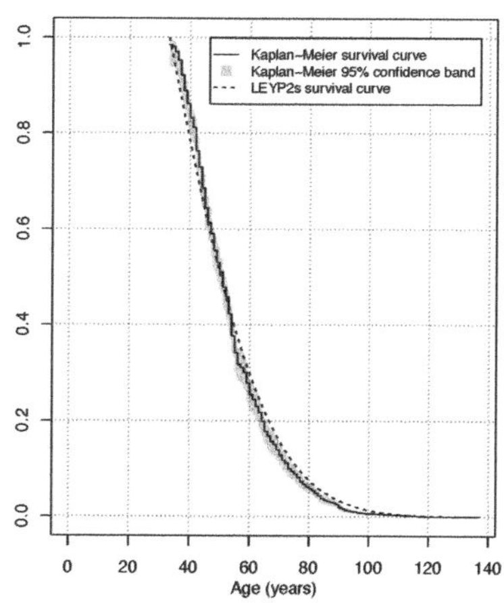


**Figure 8.13.** *Empirical versus LEYP2s survival – Lausanne gray cast iron data*

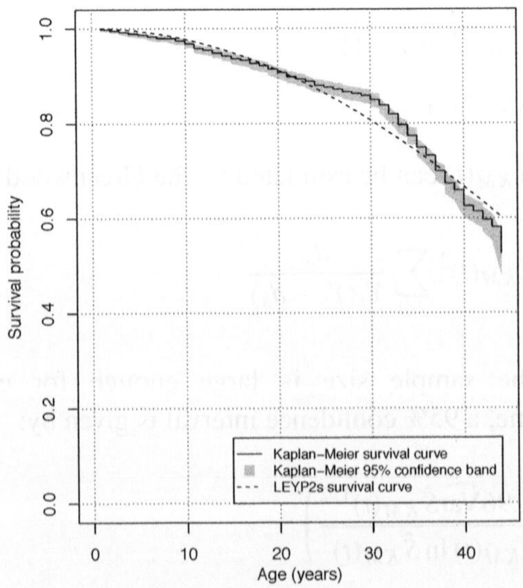


**Figure 8.14.** *Empirical versus LEYP2s survival – Lausanne ductile iron data*

# 9

# Conclusion and Outlook

Our presentation of the linear extension of the Yule process (LEYP) model, and of its adaptation to account for selective survival (LEYP2s), as well as the case study application, have mainly considered the infrastructure asset management (IAM) of water networks from a *reliability* angle. We will give in this concluding chapter some supplements concerning the software implementation of the LEYP model, and model enhancement needs aroused by experience feedbacks from water utilities. We will then envisage the application of the LEYP2s model to IAM decision, helping from an extended angle, by considering a *risk* approach.

## 9.1. Software implementation: Casses

The LEYP model has been applied since 2008 by several water utilities in Europe and North America. Putting this into practice has been greatly facilitated by the provision for free of the software *Casses*. As presented by Ranaud [REN 12], Casses does not just provide a statistical analysis tool, but also features a whole bunch of failure and segment description data exploration tools. The use of Casses proves to be beneficial not only for improving the annual renovation work programming, but also for inciting water utilities to complete and enhance their information system, thus initiating a potential virtuous circle in water utility management practices.

## 9.2. Model enhancement needs

LEYP model practical implementation has also revealed potential shortcomings, which, besides the necessity of LEYP2s enhancement, mainly pertain to the lack of flexibility of the aging function in $\lambda(t)$ and the impossibility in the current version to use time-dependent covariates.

### 9.2.1. *More flexible analytical form for the failure intensity function*

As defined in [5.4], the aging factor consists of a power function of time $\delta t^{\delta-1}$. This involves both a continuous increase in the failure intensity with age, which in some applications may be somewhat irrealistic, as emphasized by Eisenbeis [EIS 94], and also an intensity of aging independent of the characteristics of the technical objects. Two ways of improvement can then be envisaged for future research work:

– either making the $\delta$ parameter dependent on covariates;

– or adapting a more flexible aging function, different from a power function.

A different aging function could be, for example:

$$\lambda(t) = \left(1 - \exp(-t^{\delta_1} e^{\mathbf{Z}_1^{\mathrm{T}} \beta_1})\right)^{\delta_2} e^{\mathbf{Z}_2^{\mathrm{T}} \beta_2} \qquad [9.1]$$

The analytical form of [9.1], on the one hand, is much more complicated, but allows, on the other hand, to split the covariates into one set $\mathbf{Z}_1$ that modulates the aging intensity (the shape of $\lambda(t)$), and another set $\mathbf{Z}_2$ that keeps the Proportional Hazard Model (PHM) property. Moreover, this analytical form [9.1] ensures that $\lambda(t)$ does not increase indefinitely with age but rather reaches a plateau.

### 9.2.2. *Time-dependent covariates*

As shown by Kleiner-Rajani [KLE 02], water network failures can be strongly influenced by climatic effects. This phenomenon has been extensively studied by Babykina [BAB 10] (see also [BAB 14]) in the LEYP model framework. A major conclusion is that introducing time-dependent covariates, namely climate-related covariates, into the covariates vector $\mathbf{Z}$ in

[5.4] results in better LEYP parameter estimates, particularly concerning $\alpha$ and $\delta$ parameters.

A good illustration can be found in [CLA 14], related to the introduction of a frost-related covariate into the LEYP model, with an application by *Lyonnaise des Eaux* (a French water company) to the water utility network failure data for Bordeaux, France. Table 9.1 shows the LEYP parameter estimated with ductile iron and steel pipe failure data, when introducing (or not) a time-dependent frost-related covariate into the model; these results suggest that a frost-related covariate helps to better assess the effect of aging ($\delta$ is significantly increased, probably by mitigating the effect of past failures as $\alpha$ is decreased), as well as the effects of service pressure and connecting pipe density. Failure number predictions appear as a consequence to be more accurate, as shown by Figure 9.1.

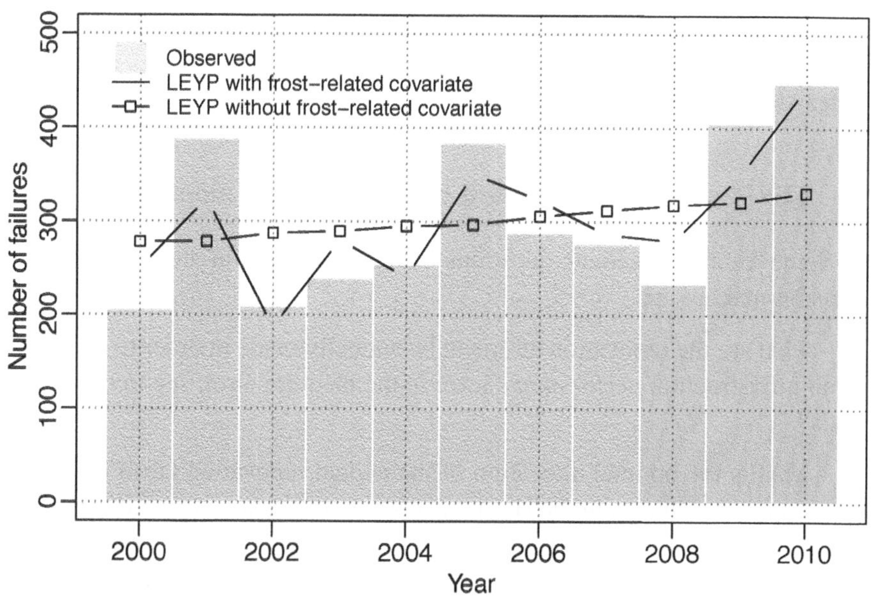


**Figure 9.1.** *Annual numbers of failure observed for all materials of Bordeaux water utility network, compared to LEYP predictions, with and without a time-dependent covariate (from [CLA 14])*

| Label | Estimates without frost-related covariate (P-value) | Estimates with frost-related covariate (P-value) |
|---|---|---|
| $\alpha$ | 6.91 (0.000) | 5.77 (0.000) |
| $\delta$ | 1.12 (0.019) | 1.25 (0.000) |
| $\beta_0$ | -3.93 (0.000) | -4.24 (0.000) |
| $\beta_1$ (lnlength) | 0.53 (0.000) | 0.55 (0.000) |
| $\beta_2$ (Pipe diameter) | -0.001 (0.740) | -0.002 (0.095) |
| $\beta_3$ (Service pressure) | -0.059 (0.740) | 0.257 (0.095) |
| $\beta_4$ (Ground corrosivity) | 0.540 (0.000) | 0.543 (0.000) |
| $\beta_5$ (Connection density) | 2.279 (0.008) | 2.676 (0.001) |
| $\beta_6$ (Frost) | | 0.070 (0.000) |

**Table 9.1.** *LEYP parameter estimates with and without time-dependent (frost-related) covariate (from [CLA 14])*

A similar introduction of time-dependent covariates into the LEYP2s model would be undoubtedly beneficial, and forthcoming research works will have to endeavor to implement this improvement.

## 9.3. LEYP2s model as element of IAM decision helping

Some central practical questions to be answered in the field of IAM decision-making are:

– what are the renovation efforts to be annually made in order to maintain a given infrastructure *performance* level in the medium term, say in the next 10 years?

– what is the optimal allocation of the annual renovation effort among the infrastructure elements in the short term, say in the next year or next 2 years?

The first question is of strategic interest, i.e. in the medium term, and arises at a global spatial scale of the infrastructure; it eventually consists of defining the annual renovation rates, and related budgets, to be implemented in the next 10 years. Whereas the second question is of tactical interest, i.e. in the short term, at the local scale of infrastructure elements, and consists of optimally spending the annual budget fixed at the aforementioned strategic level. These considerations have three important consequences:

– the predictions of the failure model have to be accurate in the short-term at the infrastructure element scale;

– the predictions have to be as unbiased as possible in the mid- or long-term at a more global scale;

– comparing IAM strategies involves to simulate infrastructure failures and renovations in the long term (i.e. beyond 50 years), and thus to couple the failure model with a service lifetime model.

### 9.3.1. *Accounting for vulnerability to failures: toward a risk approach*

From the sole reliability angle, LEYP2s provide a relevant modeling tool, which is nevertheless insufficient per se. The question of infrastructure performance is indeed pivotal in IAM, and involves considering the *vulnerability* tied to infrastructure elements, as presented by Large [LAR 15].

In the case of a water network, the failure of a given network segment entails water supply disruptions that affect users directly connected to the segment, as well as users connected to other segments, the hydraulic of which may depend on the failed segment; all users are, moreover, not equally sensitive to water supply disruption, as their economic activity (or health) may depend more or less dramatically on water supply. The failure of a pipe can also damage its structural environment (e.g. basement flooding) and neighboring infrastructures, and the subsequent repair work involves nuisances (road traffic disruption and noise).

Depending on the hydraulic importance of the failed pipe and the socioeconomic and environmental characteristics of the failed pipe neighborhood, the extent of the failure impact can be quantified, and a vulnerability measure can then be assigned to each network segment. Most advanced water utilities have such vulnerability data well and accurately formalized within their information system. This information has moreover already been used for years to build the annual renovation work planning of the water network, the renovation priorities being determined according to the failure *risk* of the network segments; the term *risk* must here be understood as the product of the failure probability and vulnerability measure.

The observed service lifetime distributions, formalized by proposition 7.2 and studied in section 8.7, are therefore likely to bear some dependency upon the segment vulnerability. A major stake of forthcoming research works will consist of making function $\psi(t)$ and parameter $\phi$ of definition 6.1 depend on vulnerability, thus making the LEYP2s model compliant with a risk-based IAM approach.

# Appendices

# Appendix A

# Product Integration

In real analysis, product integration plays a role for products similar to that played by Riemann–Stiltjes integration for sums. We will illustrate its construct and calculation through an example, which is used in Chapters 5 and 7, for a LEYP and LEYP2s likelihood calculation.

We consider to this end a time interval $[s, t[$ partitioned into $n$ subintervals, as illustrated by Figure A.1. Without loss of generality for our purpose, we consider subintervals $[t_{i-1}, t_i[, i = 1, \ldots, n$ of equal lengths $(t - s)/n$.

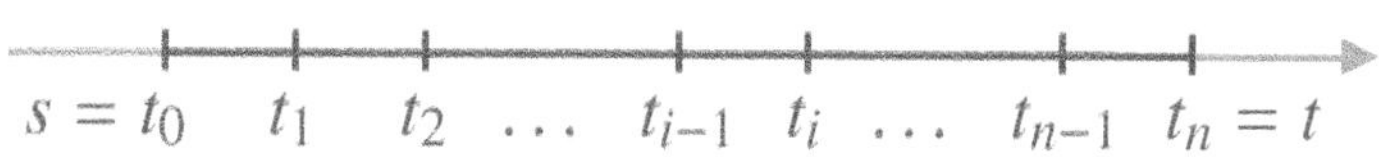


**Figure A.1.** *Partition of time interval* $[s, t[$

We consider a real function $F(t)$ which is continuously differentiable, and calculate the discrete product:

$$\prod_{i=1}^{n} (1 - (F(t_i) - F(t_{i-1})))$$

Passing to the limit for $n \to \infty$ allows us to define the following product integral:

$$\prod_{u \in [s,t]} (1 - \mathrm{d}F(u)) = \lim_{n \to \infty} \prod_{i=1}^{n} (1 - (F(t_i) - F(t_{i-1})))$$

We can now establish the following useful proposition.

PROPOSITION A.1.–

$$\prod_{u\in[s,t]} (1 - \mathrm{d}F(u)) = \exp\left(-\int_s^t \mathrm{d}F(u)\right)$$

PROOF.–

The proof is based on the following property of the ln function:

$$\lim_{x\to 0+} \ln(1 - x) = -x$$

We can then write:

$$\begin{aligned}
&\prod_{u\in[s,t]} (1 - \mathrm{d}F(u)) \\
&= \lim_{n\to\infty} \mathrm{e}^{\ln \prod_{i=1}^{n}(1-(F(t_i)-F(t_{i-1})))} \\
&= \lim_{n\to\infty} \mathrm{e}^{\sum_{i=1}^{n} \ln(1-(F(t_i)-F(t_{i-1})))} \\
&= \lim_{n\to\infty} \mathrm{e}^{-\sum_{i=1}^{n}(F(t_i)-F(t_{i-1}))} \\
&= \mathrm{e}^{-\int_s^t \mathrm{d}F(u)}
\end{aligned}$$

□

# Appendix B

# An Algebraic Identity

The following proposition states the algebraic identity used to prove proposition 3.2:

PROPOSITION B.1.–

$$-\sum_{k=0}^{m-1} \frac{1}{\prod_{l=0,l\neq k}^{m} (\alpha_l - \alpha_k)} = \frac{1}{\prod_{l=0,l\neq m}^{m} (\alpha_l - \alpha_m)} \qquad [B.1]$$

PROOF.– Let $f()$ be any function, and $D_0 = 1, D_1 = x-\alpha_0, D_2 = (x-\alpha_0)(x-\alpha_1)$, $\ldots, D_m = (x-\alpha_0)(x-\alpha_1)\ldots(x-\alpha_{m-1})$ the first $m$ Newton polynomials, the real-valued sequence $S = \{\alpha_j, j = 0, 1, \ldots, m-1\}$ being included in a bounded interval, and assumed to be increasing (without loss of generality for our purpose).

We recall the following definition of the Lagrange interpolating polynomial of $f()$ on $S$:

$$L_S(f) = \sum_{i=0}^{m} \Delta_i(\alpha_0, \ldots, \alpha_i) D_i$$

where the divided differences $\Delta_i(\alpha_0, \ldots, \alpha_i)$ of $f()$ with respect to $S$ are defined by recursion:

$$\Delta_0(\alpha_0) = f(\alpha_0)$$

$$\Delta_1(\alpha_0, \alpha_1) = \frac{f(\alpha_0) - f(\alpha_1)}{\alpha_0 - \alpha_1}$$

$$\dots$$

$$\Delta_i(\alpha_0, \dots, \alpha_i) = \frac{\Delta_{i-1}(\alpha_0, \dots, \alpha_{i-1}) - \Delta_{i-1}(\alpha_1, \dots, \alpha_i)}{\alpha_0 - \alpha_i}$$

The proof is based on the following classical result, related to divided differences:

$$\Delta_i(\alpha_0, \dots, \alpha_i) = \sum_{k=0}^{i} \frac{f(\alpha_k)}{\prod_{l=0, l \neq k}^{i} (\alpha_k - \alpha_l)} \qquad \text{[B.2]}$$

To that end, we choose $f()$ such as $f(\alpha_i) = 1, \forall i \in \{0, 1, \dots, m-1\}$. equation [B.2] becomes then at order $m$:

$$\Delta_m(\alpha_0, \dots, \alpha_m) = \sum_{k=0}^{m} \frac{1}{\prod_{l=0, l \neq k}^{m} (\alpha_k - \alpha_l)} = \sum_{k=0}^{m} \frac{(-1)^m}{\prod_{l=0, l \neq k}^{m} (\alpha_l - \alpha_k)}$$

Furthermore, except at order 0, all divided differences cancel and we find:

$$\sum_{k=0}^{m} \frac{1}{\prod_{l=0, l \neq k}^{m} (\alpha_l - \alpha_k)} = 0$$

or, equivalently:

$$-\sum_{k=0}^{m-1} \frac{1}{\prod_{l=0, l \neq k}^{m} (\alpha_l - \alpha_k)} = \frac{1}{\prod_{l=0, l \neq m}^{m} (\alpha_l - \alpha_m)}$$

□

# Bibliography

[AAL 08] AALEN O.O., BORGAN Ø., GJESSING H.K., *Survival and Event History Analysis*, 1st edition, Springer, New York, 2008.

[ABR 72] ABRAMOWITZ M., STEGUN I., *Handbook of Mathematical Functions*, 12th edition, Dover Publication Inc., New York, 1972.

[AND 93] ANDERSEN P.K., BORGAN Ø., GILL R.D. *et al.*, *Statistical Models Based on Counting Processes*, 1st edition, Springer-Verlag, New York, 1993.

[ASF 15] ASFAW Z.G., LINDQVIST B.H., "Extending minimal repair models for repairable systems: a comparison of dynamic and heterogeneous extensions of a nonhomogeneous Poisson process", *Reliability Engineering and System Safety*, vol. 140, pp. 53–58, 2015.

[BAB 10] BABYKINA G., Modélisation statistique d'événements récurrents. Exploration empirique des estimateurs, prise en compte d'une covariable temporelle et application aux défaillances des réseaux d'eau, PhD Thesis, University of Bordeaux, 2010.

[BAB 14] BABYKINA G., COUALLIER V., "Modelling pipe failures in water distribution systems: accounting for harmful repairs and a time-dependent covariate", *International Journal of Performability Engineering*, vol. 10, no. 1, pp. 41–52, 2014.

[BER 08] BERARDI L., GIUSTOLISI O., KAPELAN Z. *et al.*, "Development of pipe deterioration models for water distribution systems using EPR", *Journal of Hydroinformatics*, vol. 10, no. 2, pp. 113–126, 2008.

[BHA 97] BHARUCHA-REID A.T., *Elements of the Theory of Markov Processes and Their Applications*, Dover Publication Inc., Mineola, New York, 1997.

[BUR 10] BURN S., MARLOW D., TRAN D., "Modelling asset lifetimes and their role in asset management", *Journal of Water Supply: Research and Technology – AQUA*, vol. 59, nos. 6–7, pp. 362–377, 2010.

[CHA 02] CHANG C.C.H., CHAN W., KAPADIA A.S., "The analysis of recurrent failure times : the time-dependent Yule process approach", *Communications in Statistics–Theory and Methods*, vol. 31, no. 7, pp. 1203–1213, 2002.

[CLA 14] CLAUDIO K., LE GAT Y., COUALLIER V. *et al.* "Integration of time-dependent covariates in recurrent events modelling: application to failures on drinking water networks", *Journal de la Socièté Française de Statistique*, vol. 155, no. 3, pp. 62–77, 2014.

[COO 02] COOK R.J., LAWLESS J.F., "Analysis of repeated events", *Statistical Methods in Medical Research*, vol. 11, pp. 141–166, 2002.

[COX 62] COX D.R., *Renewal Theory*, Chapman and Hall, New York, 1962.

[COX 72] COX D.R., "Regression models and life tables (with discussion)", *Journal of the Royal Statistical Society, Series B*, vol. 34, pp. 187–220, 1972.

[COX 74] COX D.R., HINKLEY D.V., *Theoretical Statistics*, Chapman and Hall, London, 1974.

[DEB 10] DEBÒN A., CARRIÒN A., CABRERA E. *et al.*, "Comparing risk of failure models in water supply networks using ROC curves", *Reliability Engineering and System Safety*, vol. 95, pp. 43–48, 2010.

[EIS 94] EISENBEIS P., Modélisation statistique de la prévision des défaillances sur les conduites d'eau potable, PhD Thesis, University of Louis Pasteur Strasbourg, 1994.

[GRE 20] GREENWOOD M., YULE G.U., "An inquiry into the nature of frequency distributions representative of multiple happenings with particular reference to the occurrence of multiple attacks of disease or of repeated accidents", *Journal of the Royal Statistical Society, Series A*, vol. 83, pp. 255–279, 1920.

[GRE 96] GREENWOOD P.E., NIKULIN M.S., *A Guide to Chi-Squared Testing*, John Wiley and Sons, New York, 1996.

[HER 03] HERZ R.K., LIPKOW A.T., "Strategic water network rehabilitation planning" *Water Science and Technology: Water Supply*, vol. 3, nos. 1–2, pp. 35–42, 2003.

[KLE 01] KLEINER Y., RAJANI B., "Comprehensive review of structural deterioration of water mains: statistical models", *Urban Water*, vol. 3, no. 3, pp. 131–150, 2001.

[KLE 02] KLEINER Y., RAJANI B., "Forecasting variations and trends in water main breaks", *Journal of Infrastructure Systems*, vol. 8, no. 4, pp. 122–131, 2002.

[LAR 15] LARGE A., LE GAT Y., ELACHACHI S. *et al.*, "Decision support tools: review of risk models in drinking water network asset management", *Water Utility Journal*, vol. 10, pp. 45–53, 2015.

[LAW 87] Lawless J.F., "Regression methods for Poisson process data", *Journal of the American Statistical Association, Theory and Methods*, vol. 82, no. 399, pp. 808–815, 1987.

[LAW 95] Lawless J.F., Nadeau C., "Some simple robust methods for the analysis of recurrent events", *Technometrics*, vol. 37, no. 2, pp. 158–168, 1995.

[LEG 00] Le Gat Y., Eisenbeis P., "Using maintenance records to forecast failures in water networks", *Urban Water*, vol. 2, pp. 173–181, 2000.

[LEG 01] Le Gauffre P., Ruffier J., Tanguy C. *et al.*, Projet CAPTUR – Consolidation d'un cadre théorique d'analyse des patrimoines techniques urbains de type réseau, Internal report, INSA, Lyon, France, 2001.

[LEG 02] Le Gat Y., Evaluation de la performance d'un modèle de prévision des cases en réseau d'adduction d'eau potable, Thesis, University of Bordeaux 2, 2002.

[LEG 14] Le Gat Y., "Extending the Yule process to model recurrent pipe failures in water supply networks", *Urban Water Journal*, vol. 11, no. 8, pp. 617–630, 2014.

[LOR 05] Lorenz M., "Methods of measuring the concentration of wealth", *Publications of the American Statistical Association*, vol. 9, no. 70, pp. 209–219, 1905.

[MCC 89] McCullagh P., Nelder J., *Generalized Linear Models*, 2nd edition, Chapman and Hall, London, 1989.

[NEL 65] Nelder J.A., Mead R., "A simplex method for function minimization", *Computer Journal*, vol. 7, pp. 308–313, 1965.

[PEL 99] Pelletier G., Impact du remplacement des conduites d'acqueduc sur le nombre annuel de bris, PhD Thesis, University of Québec, INRS-Eau, 1999.

[PRE 02] Press W.H., Teukolsky S.A., Vetterling W.T. *et al.*, *Numerical Recipes in C – The Art of Scientific Computing*, 2nd edition, Cambridge University Press, New York, 2002.

[RD 11] R Development Core Team, R: a Language and Environment for Statistical Computing, R Foundation for Statistical Computing, Vienna, Austria, 2011.

[RAO 73] Rao C.R., *Linear Statistical Inference and its Applications*, 2nd edition, John Wiley and sons, New York, 1973.

[REN 66] Renyi A., *Calcul des probabilités*, Dunod, Paris, 1966.

[REN 12] Renaud E., Le Gat Y., Poulton M., "Using a break prediction model for drinking water networks asset management: from research to practice", *Water Science & Technology: Water Supply*, vol. 12, no. 5, pp. 674–682, 2012.

[ROS 83] Ross S., *Stochastic Processes*, John Wiley and Sons, New York, 1983.

[ROS 97] Ross S., *Simulation*, 2nd edition, Academic Press, San Diego, 1997.

[RØS 00] RØSTUM J., Statistical modelling of pipe failures in water networks, Thesis, Norwegian University of Science and Technology, Trondheim, Norway, 2000.

[SAE 05] SAEGROV S., *CARE-W Computer Aided Rehabilitation for Water Networks*, IWA Publishing, 2005.

[SAM 94] SAMSET O., Reliability estimation based on operating history of repairable systems, PhD Thesis, The Norwegian Institute of Technology, 1994.

[SEN 99] SEN A., BALAKRISHNAN N., "Convolution of geometrics and a reliability problem", *Statistics and Probability Letters*, vol. 43, pp. 421–426, 1999.

[STC 12] ST CLAIR A., SINHA S., "State-of-the-technology review on water pipe condition, deterioration and failure rate prediction models", *Urban Water Journal*, vol. 9, no. 2, pp. 85–112, 2012.

[THE 90] THERNEAU T.M., GRAMBSCH P.M., FLEMING T.R., "Martingale-based residuals for survival models", *Biometrika*, vol. 77, no. 1, pp. 147–160, 1990.

[YAM 09] YAMIJALA S., GUIKEMA S., BRUMBELOW K., "Statistical models for the analysis of water distribution system pipe break data", *Reliability Engineering and System Safety*, vol. 94, pp. 282–293, 2009.

[YUL 24] YULE G., "A mathematical theory of evolution, based on the conclusions of Dr. J.C. Willis, F.R.S.", *Philosophical Transactions of the Royal Society of London, Series B*, vol. 213, pp. 21–84, 1924.

# Index

**R, S, T, W, Y**

Other titles from

ISTE

in

Mathematics and Statistics

## 2015

DEVOLDER Pierre, JANSSEN Jacques, MANCA Raimondo
*Basic Stochastic Processes*

## 2014

COOKE Roger M., NIEBOER Daan, MISIEWICZ Jolanta
*Fat-tailed Distributions: Data, Diagnostics and Dependence*

MACKEVIČIUS Vigirdas
*Integral and Measure: From Rather Simple to Rather Complex*

PASCHOS Vangelis Th
*Combinatorial Optimization – 3-volume series – 2nd edition*
*Concepts of Combinatorial Optimization / Concepts and Fundamentals – volume 1*
*Paradigms of Combinatorial Optimization – volume 2*
*Applications of Combinatorial Optimization – volume 3*

## 2013

COUALLIER Vincent, GERVILLE-RÉACHE Léo, HUBER Catherine, LIMNIOS Nikolaos, MESBAH Mounir
*Statistical Models and Methods for Reliability and Survival Analysis*

JANSSEN Jacques, MANCA Oronzio, MANCA Raimondo
*Applied Diffusion Processes from Engineering to Finance*

SERICOLA Bruno
*Markov Chains: Theory, Algorithms and Applications*

## 2012

BOSQ Denis
*Mathematical Statistics and Stochastic Processes*

CHRISTENSEN Karl Bang, KREINER Svend, MESBAH Mounir
*Rasch Models in Health*

DEVOLDER Pierre, JANSSEN Jacques, MANCA Raimondo
*Stochastic Methods for Pension Funds*

## 2011

MACKEVIČIUS Vigirdas
*Introduction to Stochastic Analysis: Integrals and Differential Equations*

MAHJOUB Ridha
*Recent Progress in SCombinatorial Optimization – ISCO2010*

RAYNAUD Hervé, ARROW Kenneth
*Managerial Logic*

## 2010

BAGDONAVIČIUS Vilijandas, KRUOPIS Julius, NIKULIN Mikhail
*Nonparametric Tests for Censored Data*

BAGDONAVIČIUS Vilijandas, KRUOPIS Julius, NIKULIN Mikhail
*Nonparametric Tests for Complete Data*

IOSIFESCU Marius *et al.*
*Introduction to Stochastic Models*

VASSILIOU PCG
*Discrete-time Asset Pricing Models in Applied Stochastic Finance*

## 2008

ANISIMOV Vladimir
*Switching Processes in Queuing Models*

FICHE Georges, HÉBUTERNE Gérard
*Mathematics for Engineers*

HUBER Catherine, LIMNIOS Nikolaos *et al.*
*Mathematical Methods in Survival Analysis, Reliability and Quality of Life*

JANSSEN Jacques, MANCA Raimondo, VOLPE Ernesto
*Mathematical Finance*

## 2007

HARLAMOV Boris
*Continuous Semi-Markov Processes*

## 2006

CLERC Maurice
*Particle Swarm Optimization*

Printed and bound by CPI Group (UK) Ltd, Croydon, CR0 4YY

13/09/2026

14961018-0003